Practical Paranoia™
macOS 13
Security Essentials

- ☐ The Easiest
- ☑ Step-By-Step
- ☑ Most Comprehensive
- ☑ Guide To Securing Data and Communications
- ☑ On Your Home and Office macOS Computer

Marc L. Mintz, MBA-IT, ACTC, ACSP

TPP
The Practical Paranoid™

Highest Praise for
Practical Paranoia Security Essentials

★ ★ ★ ★ ★

The author claims this is for the "new to average user (not the IT Professional) in mind." I'm not convinced of that. I think there are many IT Professionals who could learn a great deal from this book. That comment is not meant to denigrate IT Professionals (I are one) but to indicate just how much valuable information is included in this book. And for the beginner or average user who continually read about the vulnerability of even the Macintosh community "hooked" to the internet, this book lays it out how to protect yourself from those less than "slugs" who want nothing better than to empty your bank account.

- Santa Fe Cowboy (Amazon Verified Purchase Review)

★ ★ ★ ★ ★

My wife's e-mail got hacked about ten days ago, and that provided the impetus to finally tackle security issues for my Mac. I know I've been living on borrowed time.

Practical Paranoia arrived two days ago, and I started to glance through it at lunch yesterday. Well, three hours later I had "glanced" through most of it. It is an easy read.

This book is exactly what I hoped for... a little bit of data to convince me one last time that I need to do this, a little anecdotal humor, and then explicit step by step instructions (screen shots) on everything I need to do to protect my computer and my data. Marc even provides "assignments" to make sure the "reader "gets it."

Not only are we instructed on ways to protect us from the evil outside, but there are also instructions to protect data from hardware malfunctions.

This is exactly the way I need to learn and do.

- Photodog (Amazon Verified Purchase)

★ ★ ★ ★ ★

I became a client of Marc's five or six years ago. So, when Marc announced his new book, Practical Paranoia, I didn't waste any time buying a copy.

The reason is simple, when Marc tells you something, you can take it to the bank.

The book is clean, well-organized, designed to be completely accessible for the non-geek, and provides step-by-step screen shots and instructions on how to save yourself from the nightmare of data loss or breach.

I can tell you from my own bitter experience with a security breach we experienced on the cloud a few years ago (not Marc's purview) that cost me more than two months of my life, an ounce of prevention is worth a *ton* of cure.

And in this book, for a few bucks, Marc has laid out the preventative roadmap that can save you more time, money, and misery than you may imagine.

Former Intel CEO Andy Grove entitled his memoirs Only the Paranoid Survive. In the wild world of connected everything, Practical Paranoia is an important guide to survival.

Buy it. Implement it. You'll sleep better.

- Lanny Goodman (Amazon Verified Purchase)

★ ★ ★ ★ ★

Easy to use - I did about half of the recommendations in under an hour. I'm a writer and I run a business from home. I'm not so technically adept. When I travel, I hate packing up all my equipment and putting things in storage to make sure that, in the unlikely event the house is robbed, all my data is safe, and I could easily restart my business from a backup in the cloud. This book is a "paint-by-numbers" approach to security for every aspect of a computer system, from backups to email to encrypting the physical drives. There's really nothing else on the subject unless you want to read technical blogs for days. This book has it all in one place and makes it easy for even a normal person (like me).

- Ben Taxy (Amazon Verified Purchase)

Copyright

Dedication

To Candace,
without whose support and encouragement
this work would not be possible

Contents at a Glance

Contents in Detail

Contents in Detail

Contents in Detail

Contents in Detail

Practical Paranoia macOS 13 Security Essentials

Marc L. Mintz, MBA-IT, ACTC, ACSP

Proof of Purchase

Practical Paranoia macOS 13 Security Essentials

Full Name: _____

Mailing Address: _____

_____

_____

Email Address: _____

Phone: _____

Educational Institution: _____

Position: _____

For what are you redeeming your Proof of Purchase? _____

_____

_____

Please attach a copy of your receipt

Mail to:

The Practical Paranoid, LLC
1000 Cordova Pl #842
Santa Fe, NM 87505

Thank You for Studying Practical Paranoia!

Dear Student,

Thank you for getting this far into this book. Although I cannot promise it will be as easy getting all the way through as it was to get here, I do promise this is the easiest and most comprehensive book in this category that you can buy.

When I wrote the first edition of Practical Paranoia, I received many emails and calls from instructors, students, and fans thanking me for the book. In truth, over half of this book came out of the questions and insights provided by the readers themselves. I love the feedback. I invite you to write to me at *marc@thepracticalparanoid.com.*

I also ask a favor. Please write a review of *Practical Paranoia*. Loved it, hated it, what worked for you, what you would like to see added or changed. I both enjoy and value your feedback.

Reviews can be difficult to come by these days. You, the reader, have the power now to make, break, and shape the evolution of a book. If you have the time, please visit my author page on Amazon.com[1]. Here you can find all my books and leave a review.

Thank you so much for studying Practical Paranoia and for spending time with me.

Warmly,

_____

[1] *https://www.amazon.com/author/marclmintz*

Chapter Length: 40 minutes

1 Introduction

Just because you're paranoid doesn't mean they aren't after you.
–Joseph Heller[1], Catch-22

Everything in life is easy–once you know the how.
–Marc L. Mintz[2]

What You Will Learn in This Chapter

- What is Cybersecurity and Internet Privacy
- Who should study this course
- What is unique about this course
- Why worry?
- Reality check
- About the author
- Practical Paranoia updates
- Notes for instructors, teachers, and professors
- Update bounty
- Format conventions used in this book
- What is new in macOS 13 cybersecurity and internet privacy

---

[1] *https://en.wikipedia.org/wiki/Joseph_Heller*
[2] *https://thepracticalparanoid.com*

What You Will Need in This Chapter

- No additional resources required.

1.1   What is Cybersecurity and Internet Privacy

Cybersecurity

Merriam-Webster dictionary defines cybersecurity:

> measures taken to protect a computer or computer system (as on the Internet) against unauthorized access or attack.

I think the *Cybersecurity & Infrastructure Security Agency (CISA)*[3] says it better:

> Cybersecurity is the art of protecting networks, devices, and data from unauthorized access or criminal use and the practice of ensuring confidentiality, integrity, and availability of information.

Internet Privacy

Wikipedia defines internet privacy:[4]

> Involves the right or mandate of personal privacy concerning the storing, repurposing, provision to third parties, and displaying of information pertaining to oneself via Internet.

Taken at the most fundamental level, cybersecurity and internet privacy are about protecting your personal information (passwords, sites visited, health conditions, finances, phone calls, emails, text messages, relationships, private photographs, video and audio recordings, public and private life, and so much more) from unauthorized access and use.

As our computers, mobile devices, and internet services are the central hub for storing and transmitting all our data, it makes sense to focus our resources on securing the Information Technology (IT) in our lives.

---

[3] *https://us-cert.cisa.gov/ncas/tips/ST04-001*
[4] *https://en.wikipedia.org/wiki/Internet_privacy*

1.2   Who Should Study This Course

Traditional business thinking holds that products should be tailored to a laser-cut market segment. Something like *18-25-year-old males, still living at their parents' home, who like to play video games, working a minimum-wage job.* Yup, we all have a pretty clear image of that market segment.

In the case of this security and privacy course, the target market segment is *all users of macOS computers.* Really! From my great-Aunt Rose who is wrestling with using her first computer, to the small business, to the Information Technology (IT) staff for major corporations and government agencies: this is for all of you.

There is little difference between *home-level security* and *military-grade security* when it comes to cybersecurity and privacy technology. Even though the military may use better security on their physical front doors (e.g., MPs with machine guns protecting the underground bunker compared to a home with a Kwikset deadbolt and a neurotic Chihuahua) the steps to secure computers and mobile devices for home and business use are almost identical for both private and military users.

The importance of data held in your device may be every bit as important as the data held by the CEO of a Fortune 500. Your data also is as vulnerable to penetration as the data of any other person, and the repercussions of that data being accessed are just as important. How would it feel if you lost passwords, financial information, travel scheduling, personal relationship discussions and photos of loved ones?

1.3   What is Unique About This Course and Book

By following the easy, illustrated, step-by-step instructions in this book, you can secure your computer to industry standards. The steps outlined here are the same steps used by my consulting organization when securing systems for hospitals, government agencies, and the military.

Practical Paranoia Security Essentials is the first comprehensive security book series written with the new to average user in mind as well as IT professionals and

STEM and Computer Science students. Hardening your cybersecurity and privacy helps your business protect the valuable information of you and your customers.

Should your work include HIPAA, SEC, or legal-related information, to be in full compliance with regulations it is likely that you need to be using the latest operating system version. Newer operating system (OS) versions have newer protections and filters.

Do not let the number of pages here threaten you. This book is the ultimate step-by-step guide for protecting your device, but dull background theory is reduced to a minimum. I include only what is necessary to grasp the need-for and how-to. All information and steps are built on current guidelines, policies and procedures, and best practices from Apple, Google, Microsoft, NIST, NSA, US-CERT, and my own 34 years as an IT consultant, developer, technician, and trainer.

The organization of this book is simple. Each chapter represents a major area of security and privacy vulnerability, along with the tasks you should do to protect your data, device, and personal identity. To review your work using this guide, use the *Security Checklist* provided at the end of this book.

To make your system as secure as possible, I recommend following the sequence provided in this guide. Bad guys seek out your weak points. Leave no obvious weakness and they move on to an easier target. There is an old joke: Two friends are camping in the forest when they see a bear charging toward them. The first friend asks the second: *How fast do you need to run to out-run a bear?* The second friend replies: *Just a little faster than you.*

Theodore Sturgeon, American science fiction author and critic, stated: *Ninety percent of everything is crap*[5]. Mintz's extrapolation of Sturgeon's Revelation is *Ninety percent of everything you have learned and think to be true is crap.*

Sturgeon was an optimist.

I have spent most of my adult life in distilling what is real and accurate from, well, Sturgeon's 90%. The organizations and workshops I have produced, and the *Practical Paranoia* book series, all spring from this pursuit. If you find any area of this workshop or book that you think should be added, expanded, improved, or changed, contact me personally with your recommendations at *marc@thepracticalparanoid.com.*

---

[5] *https://en.wikipedia.org/wiki/Sturgeon%27s_law*

1.4   Why Worry?

In terms of network, Internet, and data security, all users of computers and mobile devices must be vigilant because of the presence of malware[6], physical theft, malicious websites, and cybercriminals. Attacks on computer and mobile device users by tricksters, criminals, and governments are on a steep rise. How bad is the situation?

- Per a study from Ponemon Institute and IBM Security, the global average cost of a data breach in 2019 was $3.92 million.

- According to data from Kensington and the FBI, one laptop is stolen every 53 seconds, and over 70 million cell phones are **lost** each year. Of those millions of devices stolen or lost, only 3% ever are recovered.

- Typical email is clearly readable at dozens of points along the Internet highway on its trip to the recipient and may be read by someone you do not know.

- If you are in the USA, your government monitors your devices. The Cyber Intelligence Sharing and Protection Act (CISPA)[7] allows the government easy access to all your electronic communications. PRISM[8] allows government agencies to collect and track data on any American device. Do not feel smug if you're not in the USA, because you're likely tracked at least as vigorously.

The list goes on, but you get the point. It is not a matter of *if* your data will ever be threatened. It is only a matter of *when,* and how often attempts will be made.

1.5   Reality Check

Nothing can 100% guarantee 100% security 100% of the time. Even the White House and CIA websites and internal networks have been penetrated. We know that organized crime, as well as the governments of China, North Korea, Russia,

---

[6] *http://en.wikipedia.org/wiki/Malware*

[7] *http://en.wikipedia.org/wiki/Cyber_Intelligence_Sharing_and_Protection_Act*

[8] *http://en.wikipedia.org/wiki/PRISM_(surveillance_program)*

Great Britain, United States, Australia, and other nations have billions of dollars and tens of thousands of highly skilled security personnel on staff looking for *zero-day exploits*[9]. The zero-day exploits are vulnerabilities not yet discovered by the developer. As if this is not enough, the U.S. government influences the development and certification of most security protocols. This means that industry-standard tools used to secure our data have been found to include vulnerabilities introduced by U.S. government agencies.

With these odds against us, should we just throw up our hands and accept there is no way to ensure our privacy? Well, just because breaking into a locked home only requires a rock through a window, should we give up and not lock our doors? Of course not. We do everything we can to protect our valuables and not become victims. When leaving on vacation we lock doors, turn on motion detectors, notify the police to prompt additional patrols, and stop mail and newspaper delivery.

The same is true of our digital lives. For the very few who are targeted by the NSA, FBI, CIA, etc., there is little that can be done to completely block them from reading your email, following your chats, and recording your web browsing. But you can make it time and labor intensive for them to do so.

And you *can* protect yourself, your data, and your devices from being penetrated by criminals, pranksters, competitors, nosy people, a wackadoodle ex, as well as about the collateral damage caused by malware. By following this book, you can fully secure your data and your first device in two days, and any additional devices in a half day. This is a small price to pay for peace of mind and security.

It is imperative that you secure all points of vulnerability. Remember, penetration usually does not occur at your strong points. A home burglar avoids hacking at a steel door when a simple rock through a window gains entry. A strong password and encrypted storage by themselves do not protect data (including usernames and passwords) from hackers.

- Note: Throughout this book we provide suggestions on how to use various free and for-fee applications to help enforce your protection. We have used these applications with success, and thus feel confident in recommending them. Neither Marc L. Mintz, nor The Practical Paranoid, LLC. receive compensation for suggesting applications and we may change suggestions as our experience changes.

---

[9] *https://en.wikipedia.org/wiki/Zero-day_(computing)*

1.6   About the Author

Marc Louis Mintz is a respected IT consultant and trainer with over three decades of experience. Marc's enthusiasm, humor, and training expertise were honed on leading edge work in motivation, management development, and technology. His software and hardware workshops are highly rated by seminar providers, meeting planners, managers, and participants because he empowers participants to see with new eyes, think in a new light, and problem solve using new strategies. Marc holds a Master of Business Administration with specialization in IT (MBA-IT), Chauncy Technical Trainer certification, Post-Secondary Education credentials, and over a dozen industry certifications.

When away from the podium, Marc is in the trenches, working to keep his clients' systems secure and private. Well, not so much anymore. As of 2020 Marc retired as CEO and CIO of Mintz InfoTech, Inc. to spend more time with his passions: the world's best partner and dogs, Voice Over Acting, and maintenance of the very best cybersecurity and internet privacy guides.

The author may be reached at:

Marc L. Mintz
The Practical Paranoid LLC.
1000 Cordova Pl #842
Santa Fe, NM 87505
Email: *marc@thepracticalparanoid.com*
Web: *https://thepracticalparanoid.com*

1.7   Practical Paranoia Updates

Information regarding IT security changes daily, so we offer many options to keep you on top of everything.

1.7.1 Practical Paranoia Paperback Book Updates and Upgrades

We are constantly updating *Practical Paranoia* so that you have the latest, most accurate resource available. If at any time you wish to upgrade to the latest version of *Practical Paranoia* at the lowest price we can offer:

1. Tear off the front cover of your ***Practical Paranoia***.

2. Make your check payable to *The Practical Paranoid* for $32.50.

3. Send front cover, check, and mailing information to:
   The Practical Paranoid, LLC
   1000 Cordova Pl #842
   Santa Fe, NM 87505

4. Your new copy of Practical Paranoia is sent by USPS. Please allow up to four weeks for delivery.

1.7.2 Practical Paranoia Kindle Updates and Upgrades

We are constantly updating *Practical Paranoia* so that you have the latest, most accurate resource available. Unfortunately, we have no control over Kindle version upgrades (such as from Practical Paranoia macOS 10.14 to Practical Paranoia macOS 13), and these require a new purchase. But if at any time you wish to update to the latest Kindle version of *Practical Paranoia macOS 13* at no cost:

1. Delete the copy of *Practical Paranoia* that is installed on your Kindle device.

2. Download the current edition of *Practical Paranoia* from your Kindle library.

1.7.3 Practical Paranoia *Live!* Online Updates

If you have the Practical Paranoia *Live!* Online edition, you always have the latest update for your OS.

1. Launch your web browser.

2. Go to your Google Drive portal (*https://drive.google.com*) > *Shared With Me*.

3. The new version is found in your *Practical Paranoia Textbook* folder.

1.8   Note for Trainers

The *Live!* online edition is a streaming pdf version of the book. We update as the technology changes, often every month or two. We strongly recommend opting for the *Live!* online digital version of the Practical Paranoia books. These allow the students and teacher to always have the most current version—at a lower price than either the paperback or kindle versions.

Please contact our office for details:

Email: *info@thepracticalparanoid.com*
Voice: +1 505.453.0479

1.9   Update Bounty

Although we work tirelessly to keep the contents of this workbook up to date, every now then something slips by. If you discover anything in this workbook that does not reflect current reality, and you are the first to report it to us, we thank you, and you will receive a free signed copy of *Practical Paranoia.*

To make an update bounty report:

Email: *info@thepracticalparanoid.com*

1.10   [Windows] Running Programs on Windows

1.11   Format Conventions Used in this Book

Italics are used to represent hyperlinks, action steps, and file and folder locations.

Courier is used to represent user input in a command-line environment.

• **Warning** is displayed when an action is potentially destructive to your data.

- Prerequisite is displayed when additional resources are required to complete an assignment.

Footnotes with hyperlinks are displayed to provide either additional, deeper, or original source information.

[Android, iOS, Chrome OS, or Windows] in a heading indicates a chapter, section, or assignment that is used in one of the other books in the series, but not for this book or OS.

[Optional] in an assignment heading indicates an assignment that is optional for the course, typically due to the need to purchase additional resources

1 Titles are chapter headings.

1.1 Sections are section headings.

1.1.1 Assignments are assignment headings.

1.12 Threats and Vulnerabilities

Your cybersecurity and internet privacy are threatened from dozens of points of vulnerabilities–far more than even the professionals are aware. This is because in this cat-and-mouse game, there are an unknown number of unknown points of attack. These are called zero-day vulnerabilities.

Putting aside zero-day issues for the moment, let's examine the more popular vulnerabilities each of us face daily:

Your Computer, Tablet, and Phone

- *Weak or non-existent passwords.* Should a bad actor get their hands on your device with a weak password, they now have the keys to your kingdom and access to your credit cards, bank accounts, everyday communications, and dark secrets.

- *Passwords easily bypassed.* Should your device lack strong encryption, even if it has a strong password, the password can simply be bypassed.

- *Reusing Passwords.* Because it is so difficult to remember strong passwords, most people use the same few passwords for every site and service. But if one of these accounts is hacked, the criminals now have your credentials which may work on dozens of sites.

- *Malware.* Malware can gain a foothold to your device from email, website, file sharing, text message, and application. There have even been a few incidents where a legitimate antivirus application downloaded from the developer was the carrier of malware.

- *Fingerprinting.* Your device, with its combination of OS version, device model, installed fonts, memory, etc. is unique, giving it a unique fingerprint. This fingerprint is used by websites and trackers to follow your internet activities. Following your activities generates a precise profile of your likes, dislikes, political, psychological, financial, even sexual preferences.

Wi-Fi, Ethernet, and Cellular Connection Router

The first step your device makes out to the local area network or internet is through the local network transport.

- *Ethernet.* While sending data through an ethernet cable, the cable acts as a broadcast antenna. Off-the-shelf receiving devices can pick up the broadcasts, reading all your traffic from miles away[10].

- *Wi-Fi.* Most Wi-Fi networks are either unencrypted or are using encryption protocols that have been cracked. This leaves all your communications not much more private than sending smoke signals.

- *Router Password.* Most routers still use the factory default administrator username and password. This allows a bad actor administrative access to your entire network.

- *Out of Date Router Firmware.* Just as with the software on your device, the firmware on a router must be updated to patch security holes. But very few are ever updated.

- *Cellular.* Although all cellular signals are encrypted, the encryption protocol was cracked long ago.

---

[10] *https://thehackernews.com/2021/10/creating-wireless-signals-with-ethernet.html*

Internet Routers

Your communications leave your router then pass-through internet routers on the way to your Internet Service Provider (ISP). First stop is the Domain Name Service (DNS) hosted at your ISP.

- *Internet Routers.* Your internet activity can be monitored and recorded at any of the dozens of internet routers that pass along your communications.

- *DNS.* This service records each server you visit, adding to your personal profile. It is common practice for your profile to be sold to marketing groups.

Web Server

- *Monitoring and Fingerprinting.* Most websites monitor all visitors to an extreme degree. Combine this with fingerprinting and passing profile information from site to site, these organizations now literally know more about you than your mother.

- *Email.* Although you may know that your email service is encrypted, do you know if your recipient can say the same? Email containing sensitive personal or proprietary business information can easily be harvested.

- *Hacked Accounts.* Even if you use a strong password for a website, what if the site does not securely store your credentials and are then hacked. It is now common for hundreds of millions of accounts to be compromised at once.

Voice, Video, and Instant Messaging

- *Insecure Communications Software.* Voice, video, and instant messaging dominate our digital communications. But almost all the applications used store, monitor, and monetize your communications.

Social Media

- Please, do not get me started on this. It's just too early in the course.

1.13   What is at Stake

In my workshops there are always a few participants who proclaim, "I have nothing to hide, I don't really care about cybersecurity and internet privacy".

My universal reply is, "would you please write the following on the whiteboard:

- Your full legal name
- Your residential address
- Your phone number
- Your Social Security Number
- Your credit scores
- Your income
- Your computer and phone passwords
- Your passwords for Facebook, LinkedIn, Google, Microsoft, and Apple
- Your log in credentials for your bank and credit cards"

Nobody yet has taken me up on the challenge. Not because we have something to hide (although some of you do – you know who you are), but because if this information fell into the wrong hands, it would be damaging to you.

Cybersecurity and internet privacy are critically important because your finances, your relationships, your career, your company, and perhaps your life are dependent upon keeping some information out of the hands of bad actors.

The good news is it is easy to do once you know how, and we are here to guide you step-by-step.

1.14   OS Version

The world of cybersecurity and internet privacy is a 24/7/365 cat-and-mouse game. Criminals, including state agencies, have massive time and labor resources to spend discovering vulnerabilities to systems.

As we will discover later in this course, one of the primary reasons for system and app updates and upgrades is to patch security and privacy vulnerabilities. Simply put, if your system and apps are not current, there are far more opportunities for penetration than if all are current.

1.14.1 Assignment: Verify OS Version

In this assignment, you verify your macOS version. Later in the course you will learn how to set automatic system and application updates.

Check Current System Version

1. Select *Apple* menu > *About this Mac.*

2. Just under the tool bar will be the name of the current system, with the version number shown below:

3. Make note of your current system version, then close the window.

Update System

4. Tap *Apple* menu > *System Settings* > *General* > *Software Update* > *Automatic updates* more info icon.

5. Set *Automatic Updates* to *On*.

6. Enable all options, then tap *Done*:

Automatically

Check for updates	◯
Download new updates when available	◯
Install macOS updates	◯
Install application updates from the App Store	◯
Install Security Responses and system files	◯

Done

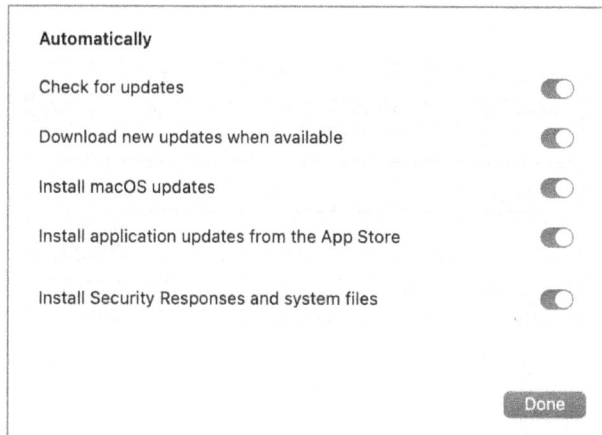

7. If any updates are available, they display. If no updates are available, the *Software Update* preference will display *Your Mac is up to date.*

8. If updates are available, tap the button to start any downloads and installs. Follow any onscreen instructions to complete installation.

9. Tap *Apple* menu > *About This Mac.* The *about* window opens to display the installed system version.

10. If the *Version* is at least *macOS 13,* you are good to go. If the *Version* is lower than *13,* your device is not designed for the current OS. It is time to replace the device.

- NOTE: Apple generally maintains security and application compatibility updates for at least the previous OS version. This means that if you are unable to update to macOS 13, you are very likely running a secure OS if you are running macOS 12. However, if you must comply with security protocols of HIPAA, SEC, NIST or other standards, you may not be in compliance unless using the latest available OS.

1.15 Legal

Cybersecurity and internet privacy is a non-stop cat-and-mouse game. If you have your devices and systems configured to the best of industry standards today, it is entirely possible that you will have a vulnerability – and perhaps a breach – tomorrow.

In addition to all the sources of vulnerabilities listed within this course, we also face:

- Software updates changing a security setting or creating a zero-day vulnerability.

- System updates changing a security setting or creating a zero-day vulnerability.

- New malware entering your device that is not recognized by your anti-malware software.

- A bug in software or system.

- Human error (you or someone using your device).

- Someone (friend, family, guest) accessing your network via their own compromised device.

The Practical Paranoid (TPP) and the authors have endeavored to provide the most effective solutions based on industry best practices to cybersecurity and internet privacy for the typical individual, household, and small to medium sized business. However, neither TPP, its agents, or the authors are your full-time cybersecurity and internet privacy consultants. By keeping this book or participating in this course, you agree to hold harmless TPP, its agents, and authors from any damages, real or otherwise.

1.16 What Is New in macOS 13 Cybersecurity and Internet Privacy

Cybersecurity and Internet Privacy issues that are new or significantly changed from macOS 12 to macOS 13 include:

- DNSSEC:
  - o Domain Name Security (DNSSEC) provides built-in DNS security, ensuring apps only talk to who they are designed to be talking to.

- Gatekeeper:
  - Previous versions only checked for malicious code during the initial launch of an app. As of macOS 13 apps and executables are checked upon every launch.

- Family Sharing:
  - Create accounts for children with parental controls.

- Home:
  - Home app displays all your smart home devices in the Home tab.

- Mail:
  - You can unsend a recent email.

- Messages:
  - You can now edit a message you just sent.
  - You can unsend a recent message.

- Passkeys:
  - Passkeys are intended to replace passwords.
  - Passkeys are a form of public-private key encryption, with a private key held securely on the user's device and a public key on the server or website.

- Private Access Tokens:
  - Intended to eliminate the need for CAPTCHAS (once implemented within websites).

- System Settings:
  - Lockdown Mode provides extreme system security for those users who may be the subject of a targeted attack.
  - System Preferences has been fully redesigned as System Settings.
  - LaunchAgents, LaunchDaemons, and Login Items can be managed from a single location in System Settings.
  - When an app adds a LaunchAgent, LaunchDaemon, or Login Item, a notification alert is displayed.

- Wallet:
  - Wallet can share your keys securely with messaging apps, including Messages, Mail, and WhatsApp.
  - You control when and where your keys can be used, and you can revoke them anytime.
  - Wallet can be used to present your ID within apps to verify your identity.

1.17   Cybersecurity and Privacy Lessons Learned

- ☐ Cybersecurity and Internet Privacy are *everyone's* business, not just for medical, banks, and ISPs.
- ☐ Maintaining cybersecurity and privacy are important to everybody who has a personal life and anything of value.
- ☐ There are very serious reasons to ensure security and privacy because you never know who may track everything you do and give access to others.
- ☐ CISPA allows the US government easy access to all your electronic communications.
- ☐ PRISM allows the US government to collect and track data on any American device.
- ☐ Zero-day exploits are vulnerabilities not yet discovered by the developer, so your malware detector may not save you.
- ☐ It does not require extensive training and expertise to protect yourself. Everything you need to know is in this guide.
- ☐ Keeping system and apps up to date is an important step to ensuring cybersecurity and internet privacy.
- ☐ New to macOS 13 is Lockdown Mode, which provides extreme security for users who may be subject to targeted attack.

2 Data Loss

Weather forecast for tonight: Dark.

–George Carlin[1]

I know, you want to jump right into cyber security and harden your awesome device. Sorry to be a Debbie Downer[2], but there is a risk of losing data in the process of some of the work ahead of us. Because of this, we must begin our exciting journey into the heart of security with drudgery: backing up your computer.

What You Will Learn in This Chapter

- Appreciate your need for backups
- Format the backup drive
- Configure Time Machine
- Integrity test Time Machine
- Create an internet backup with Google Drive

What You Will Need in This Chapter

- 1 or 2 external storage devices for local and off-site backup, 2-4 times the size of your total data.
- [Optional] iCloud+ with storage space adequate for backing up your Desktop and Documents folders. $1/month or higher.

---

[1] *https://en.wikipedia.org/wiki/George_Carlin*
[2] *https://en.wikipedia.org/wiki/Debbie_Downer*

2.1   The Need for Backups

Data loss is a very real fact of life. It is not a matter of *if* you will experience data loss, just a matter of when, and how often. Only a small percentage of computer users back up on a regular basis. I suspect these are the folks who have experienced catastrophic data loss and never want a repeat.

There are many sources of data loss. The top contenders include:

- Device theft
- Power surges
- Power sags
- Sabotage
- Fire
- Water damage. I personally have had 3 clients who have lost computers due to cats or dogs marking their territory, and my own cat took out a $4,000 monitor with nothing more than a hairball.
- Entropy / aging of the hardware
- Malware
- Terrorist activities
- Criminal activities
- Static electricity
- Physical shock to the device (banging the device, dropping, etc.)

Best practice[3] calls for at least two backups. Three are better:

- **One full backup onsite**. This allows for almost immediate recovery of lost or corrupted documents, or full recovery of the OS, applications, and documents in the event of catastrophic loss.

---

[3] *https://en.wikipedia.org/wiki/Best_practice*

- **One full backup offsite**. This is your *Plan B* in the event of a catastrophic loss of both the device and the onsite backup. This typically takes the form of fire or theft.

- **One Internet-based backup**. This is your OMG, what do I do now? fallback. Many people substitute the Internet backup for the offsite. A potential problem is your Internet backup may take days to weeks to download.

Onsite Full Backup

macOS comes with one of the most advanced backup software for any computer– Time Machine. Time Machine has several advantages over other options:

- Free

- Reliable and stable

- Low resource requirements

- Maintains document versioning. With each run, Time Machine backs up the latest version of your documents, while maintaining all prior versions as well

- Runs in the background every hour without user intervention

- Works with Migration Assistant to replicate the last backup to another Macintosh

- Creates an encrypted backup to a locally attached (USB, FireWire, or Thunderbolt) drive

As a rule, the backup drive should be at least double the size of your data, preferably quadruple. This allows for future growth and the maintenance of long-term document versioning.

A quick way to determine how much space your data uses is:

1. Tap the icon of your drive on your Desktop.

2. Select the *File > Get Info* menu.

3. Under the *General* heading, find the *Used* section. This displays how much space you currently use (which you need to protect with backup drive of at least double the space).

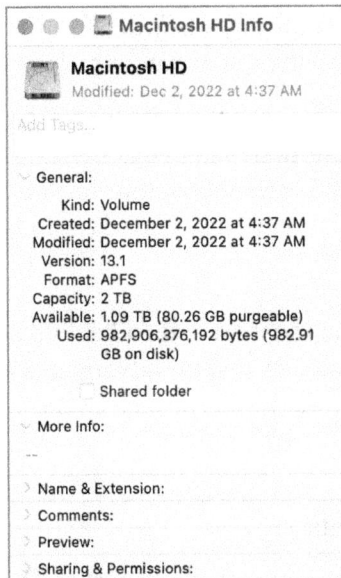

Offsite Full Backup

An offsite full backup contains the same content as the onsite backup, but after the backup completes, the backup storage device is stored offsite. Offsite backups should be performed at least once per week, preferably daily.

A good choice for offsite location is a bank safety deposit box, secured warehouse, or a trusted friends' home. The idea is to have easy access to a full backup of your computer in case a disaster like a fire or robbery leaves you without your onsite backup. This option allows much faster recovery than the cloud backup option at the expense of being slightly out of date. The average user does not create lots of data in such a short time frame, so in case of disaster you can easily grab your most recent changes from the online backup.

Internet-Based Backup

There are unique advantages to Internet-based backups:

- Disaster recovery. If a black hole opens devouring your computer, backup and offsite backup, your Internet backup always is waiting for you even after a fire destroys your home or business, or terrorist activity prevents access to the computer or offsite location.

- Remote access. Should you find yourself far away from your computer, you can access your data from any computer.

- Shared access. Some Internet-based options also allow you to share access to any documents that have been backed up.

When looking for Internet-based backup service, keep an eye out for document versioning in addition to cost, features, company, and software stability. You want the service to keep at least one month of your document versions. If a document becomes is deleted, you have a current version and a previous version available on the server.

There are downsides to Internet-based backups, including:

- Long upload and download times. It typically takes days to weeks to months to initially back up all your data, and days to download. Should a catastrophic data loss occur, and you need all drive data downloaded, the wait could be too long for your needs.

- Not all data is backed up. Unlike Time Machine, Internet-based backup does not back up your storage devices. Instead, just user data is backed up. *Applications, Library, ~/Library*, and operating system are not included in the backup. This becomes a serious issue when backing up a server, as many server processes store their databases in the Library. If this is your case, you need to create a service or script that regularly backs up such data to a secure location on the drive that can be backed up to the Internet.

My personal favorite Internet-based backup choices are:

Backblaze[4]. Easy to use, very fast uploads, full data encryption prior to upload, rock solid stable, 30-day document versioning, backs up all user accounts. For home and business. Can meet your HIPAA and SEC compliance needs.

Carbonite[5]. Versions for personal or business. Fast uploads, rock solid stable, limited document versioning, backs up all user accounts. 30-day document versioning, family and business accounts make it easier to administer multiple computers. Can meet your HIPAA or SEC compliance needs.

CrashPlan Pro[6]. For business. Fast upload, rock solid stable, document versioning, lifetime document versioning, individual and business accounts. Can meet your HIPAA or SEC compliance needs.

---

[4] *http://www.backblaze.com*
[5] *http://www.carbonite.com*
[6] *http://www.crashplan.com*

Google Drive[7] A hybrid solution. In addition to providing cloud storage and file sharing, extensions such as *Spinbackup*[8] provide a cloud-based backup of your cloud-based storage. Google *Workspace* includes data retention and eDiscovery with *Vault*[9]. Free for up to 15 GB, the paid Google Drive version can meet your HIPAA or SEC compliance needs. It includes either 1 TB or unlimited storage.

Microsoft OneDrive[10]. Like Google Drive in that it provides cloud storage and file sharing, with extensions such as *Spinbackup* providing cloud-based backup of your cloud-based storage. A unique advantage of OneDrive is the ability to have multiple people concurrently collaborate on Microsoft Office files.

SpiderOak[11] Provides both residential and business backup, end-to-end encryption, supports multiple devices, point-in-time restore, file sharing, and is HIPAA-compliant.

Backup Storage Device Format

When you buy a storage device that says it's designed for Windows, that just means it has been formatted using NTFS or FAT32–the two primary formats that Windows recognizes. In your search for the perfect backup storage device, you do not need to worry what operating system the drive claims to support. Just get the fastest one with the connection style appropriate for your computer or mobile device (SD card, USB, or USB-C) and you are good to go. The additional price for speed is more than compensated when the faster backup device recovers data 4-10 times faster than does a slower one.

Before using a storage device, it is wise to format it. This is to ensure the device is healthy, without defect, and in the proper format for your computer or mobile device.

Verify Backup Integrity Monthly

It is likely your backup system will fail from time to time. This is one reason to have multiple backups. It is wise to verify the integrity of your onsite, offsite, and

---

[7] *https://www.google.com/drive/download/backup-and-sync/*

[8] *http://spinbackup.com*

[9] *https://workspace.google.com/products/vault/*

[10] *https://www.microsoft.com/en-us/microsoft-365/onedrive/online-cloud-storage*

[11] *https://spideroak.com*

internet backups at least monthly to ensure your back up data is there when needed.

2.1.1   Assignment: Format the Backup Drive

In this assignment, you format a drive for use with Time Machine. If you follow my approach and want two backups, repeat this process with each of two drives.

- Prerequisite: One or two external storage devices, at least two-four times the capacity of your data.

- **Warning**: This assignment erases any data on the external storage device. There is no recovery.

1. Purchase an external hard drive that has at least two-four times the capacity of the data to be held on the host computer. Purchase two or more if you also perform offsite backups.

2. Change volume format to APFS.[12] Why? It is likely the drive you purchased is in FAT or NTFS format. For Time Machine use, change the format to APFS for macOS 13. If you are using an earlier version of macOS, please review the Apple Support article[13] for the storage device format appropriate for your system.

3. Connect the new drive to your computer.

4. Open *Disk Utility* located in your */Applications/Utilities* folder.

5. Select drive name from the sidebar (if you do not see a sidebar, open the width of the window).

6. Select the *Erase* button in the tool bar. The *Erase* window opens.

---

[12] *https://developer.apple.com/documentation/foundation/file_system/ about_apple_file_system*

[13] *https://support.apple.com/guide/mac-help/types-of-disks-you-can-use-with-time-machine-mh15139*

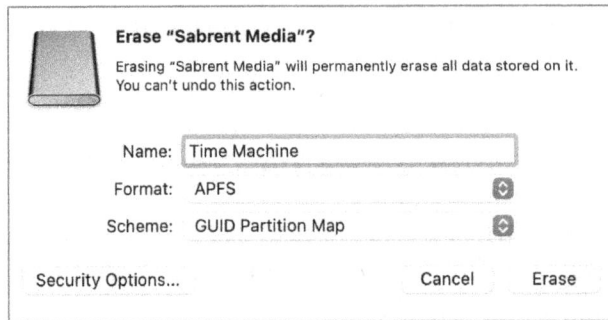

Erase "Sabrent Media"?

Erasing "Sabrent Media" will permanently erase all data stored on it. You can't undo this action.

Name: Time Machine

Format: APFS

Scheme: GUID Partition Map

Security Options...   Cancel   Erase

- In the *Name* field, enter the name you want displayed for this drive.

- In the *Format* field, select *APFS,* then select the *Erase* button.

- If the *Scheme* field is available, select *GUID Partition Map.*

7. In the *Disk Utility* window, select the *Done* button.

2.1.2   Assignment: Backup with Time Machine

Although Time Machine is designed to auto-configure, that does not mean it has auto-configured the way we want it.

In this assignment, you configure Time Machine to back up to a drive formatted in the previous assignment.

- Prerequisite: Completion of the previous assignment.

1. Attach your Time Machine drive. If you have followed the steps above, it is already attached and mounted.

2. Open *Apple* menu > *System Settings* > *General* > *Time Machine.*

3. Select *Add Backup Disk...,* then tap *Set Up Disk...*

4. Select the destination storage device. The configuration page opens.

5. In the *Maximum Space Used for Backups* area, configure to your needs.

    - Note: New to macOS 13 is the ability to automatically set aside a portion of your backup drive to use for normal storage. We strongly discourage using this option, as it results in this data being backed up to the same physical drive. Loss or damage to the Time Machine drive will lose both the data and backup.

6. Enable *Encrypt Backup,* then enter a strong password.

7. When complete, tap *Done.*

8. In *System Settings,* go back to the opening main view, select *Control Center > Time Machine,* then select *Show in menu bar.*

9. Exit System Settings.

Time Machine automatically starts to back up to this drive within the hour, then back up every hour. As storage space permits, Time Machine maintains hourly backups for the past 24 hours, daily backups for the past month, and weekly backups for all previous months.

2.1.3   Assignment: Integrity Test the Time Machine Backup

Murphy lives in technology. If anything can go wrong, it tends to do so. No different for backups.

In this assignment, you verify the Time Machine backup.

- Prerequisite: Completion of at least one Time Machine backup.

1. From the Finder menu, tap the *Time Machine* menu icon.

    - If the menu says *Backing Up: X% done - X MB copied,* Time Machine is in the process of backing up.

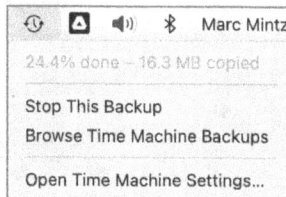

    - If the menu says *Latest Backup*... and reports a date/time within the past couple of hours, Time Machine is current.

2.1.4 [Optional] Assignment: Restore Computer from Time Machine

In the event you need to restore your computer or have a new computer you need to transfer all data to and have a Time Machine backup, *Migration Assistant* utility can restore everything from your latest or previous backups.

In this assignment, you fully restore your computer.

- **Warning**: Only perform this assignment if you need to restore your computer and you have a current Time Machine backup.

- Note: Depending on the amount of data to be transferred and the speed of the computer and storage devices, restore may take hours or a day to complete.

1. If you need to reinstall macOS due to suspected system corruption, follow instructions at *https://support.apple.com/en-us/HT204904*.

2. With your mac turned off, connect your Time Machine backup.

3. Open *Migration Assistant,* located in */Applications/Utilities*.

4. At the Migration Assistant welcome screen, select how you want to transfer your information:

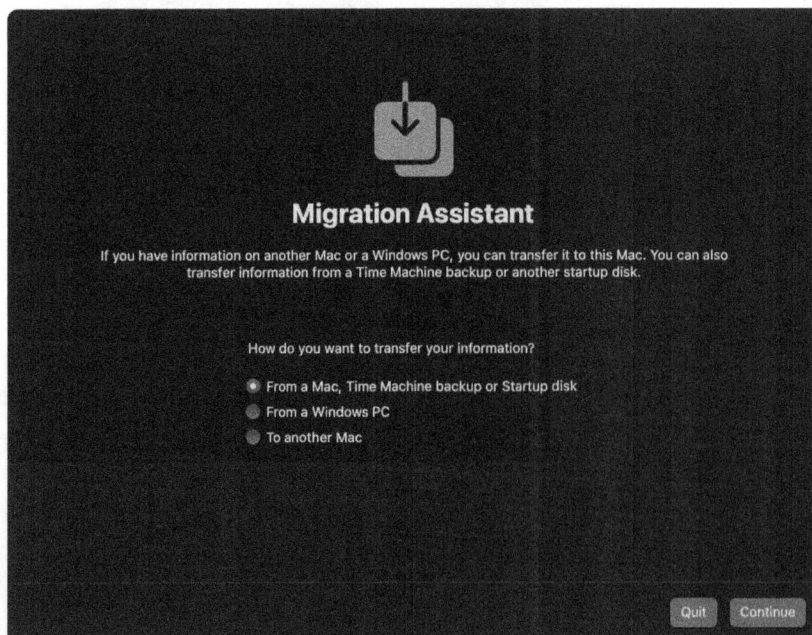

5. In the *Transfer information to this Mac screen*, select your Time Machine device, then tap *Continue*.

6. Select the target backup from within the Time Machine device, then tap *Continue*.

7. Select the information to transfer, then tap *Continue*.

8. To start the data transfer tap *Continue*.

2.1.5 [Android] Assignment: Smart Switch Backup to Local Storage

2.1.6 [Android] Assignment: Smart Switch Restore from Local Storage

2.1.7 [Windows] Assignment: Encrypt File History Backup

2.1.8 Assignment: Internet Backup with iCloud

Every Apple ID comes with 5 GB of free storage on iCloud. Additional storage space can be purchased. MacOS ships with free built-in backup software. When using iCloud backup, you can have the Desktop and Documents folders of all your macOS and Windows computers, as well as your iOS devices, synchronized with iCloud. This makes iCloud backup a good solution for some individuals.

Due to the lack of HIPAA, SEC, and NIST compliance, we don't recommend iCloud for business backup.

In this assignment you configure your macOS computer to use iCloud backup.

1. Open *Apple* menu > *System Settings* > *Apple ID* > *iCloud*.

2. Enable *iCloud Drive*.

3. Tap the *iCloud Drive Options* button.

4. Enable *Desktop & Documents Folders,* and all application check boxes.

5. Tap *Done* button.

Files and folders located within your Desktop and Documents folders will start to migrate to iCloud. When migration completes, you can access these items from the *iCloud* icon located in a Finder window sidebar.

iOS devices that are signed in with your same Apple ID and have *Settings > Apple ID > iCloud > iCloud Drive* enabled, will have access to your macOS files from the *Files* app.

2.1.9   [Optional] Assignment: Integrity Test iCloud Backup

Unfortunately, there is no integrity test that can be performed to validate an iCloud backup. But there is a quick and easy way to verify it is backing up your files.

In this assignment you will verify iCloud is backing up your files.

- Prerequisite: Completion of the previous assignment.

1. Create a new file, saving it to your Desktop or Documents folder, with a distinctive name, such as *iCloud backup test file.doc.*

2. After the file is saved, open a browser to *icloud.com.*

3. In *icloud.com,* open the *iCloud Drive* icon, and navigate to open either the *Desktop* or *Documents* folder.

4. If you find your file, iCloud can be assumed to be working.

2.1.10   [Optional] Assignment: Restore from iCloud

If you delete a file from your computer while iCloud backup is in use, iCloud retains deleted files for 30 days. Files can be immediately and permanently removed from iCloud by visiting *https://icloud.com.*

In this assignment your recover a file recently deleted from your computer and backed up with iCloud.

- Prerequisite: Completion of iCloud backup.

Recover Deleted Files Created with Apple Applications

1. Open a browser to *https://icloud.com.*

2. Tap *Recently Deleted* in the bottom-right corner.

3. Select the file(s) you wish to recover, then tap *Recover.*

Recover Files Deleted From 3rd-Party Applications

4. Open a browser to *https://icloud.com.*

5. Select *Your Name > Account Settings,* then scroll to *Advanced.*

6. In the *Advanced* field, select the type of files to be restored.

7. Select the file(s) to be restored, then tap *Restore.*

2.1.11 [Android] Assignment: Backup to a Computer with Dr. Fone

2.1.12 [Android] Assignment: Integrity Test Backup to a Computer with Dr. Fone

2.1.13 [Android] Assignment: Restore Device from Dr. Fone

2.1.14 [Windows] Assignment: Encrypt File History Backup.

2.1.15 [Optional] Assignment: Internet Backup with Google Drive

Google offers internet-based backup that is encrypted in transit (uploaded from your device to Google and downloaded from Google to your device), encrypted at rest (on the Google server), point in time recoverable, cross-platform, and includes file sharing. and with a business-class account will sign a Business Associate Agreement to be HIPAA-compliant.

In this assignment, you create a free trial Google account and then back up your data.

- The free version is limited to 15GB. Additional storage may be purchased, upgrading to the Google One license.

- The Google One[14] version costs $1.99-$9.99/month for 100GB-2TB storage, the option to add up to 5 family members, and includes chat and phone access to Google support.
- Business-Class[15] license starts at $4.20-$18.00/month for 30GB-5TB storage, includes chat and phone access to Google support and additional security options.

What Is Backed Up to Google	What Is NOT Backed Up
Apps	Bluetooth pairings
Calendars	Face ID
Chrome browser bookmarks, Smart Lock passwords	Fingerprints
	Google Authenticator
Contacts	Lock screen passwords
Docs	Notification settings
Email	SMS messages
Hangouts chat logs	
Other purchased content	
Some 3rd-party app data (verify with the developer)	
Photos	
Some system settings	

Create a Google Account

If you already have a Google Account, skip this section and then pickup on step 9.

1. Open a browser. Then *visit https://accounts.google.com.*

2. In the *Create your Google Account* page, follow the on-screen instructions to create your Google account.

---

[14] *https://one.google.com/about*
[15] *https://workspace.google.com/intl/en_id/business/*

3. Done! You may now use your Google account to set up your cloud backup.

Install Google Drive

4. Open your browser to *https://www.google.com/drive/download/*

5. In the site menu select *Individuals > Download.*

6. In the *Google Drive Terms of Service* window, tap *Agree and Download*, then tap *Download Drive for desktop* button. Google Drive downloads to your device.

7. Once download completes, open the Google Drive installer package. This mounts a volume on your Desktop.

8. Open the mounted volume. Launch the Google Drive installer, then follow the on-screen instructions to fully install Google Drive onto your computer.

Configure Google Drive

9. In your Applications folder, launch *Google Drive.*

10. Enter your Google account and password. Then tap *Connect.* Google Drive starts to synchronize with your cloud-based Google Drive.

11. By default, Google Drive creates a new folder *Google Drive* in your Home Folder. Any items placed in this folder are synchronized with the cloud-based Google Drive.

12. Google Drive defaults to synchronizing your Desktop, Documents, and Pictures folders with the cloud-based Google Drive. In the Google Drive Preferences, you can edit the folders that synchronize.

13. From the Google Drive Menu bar, tap *Gear* icon > *Preferences.*

14. From the sidebar, tap *Google Drive.* The *My Drive Synching Options* window appears:

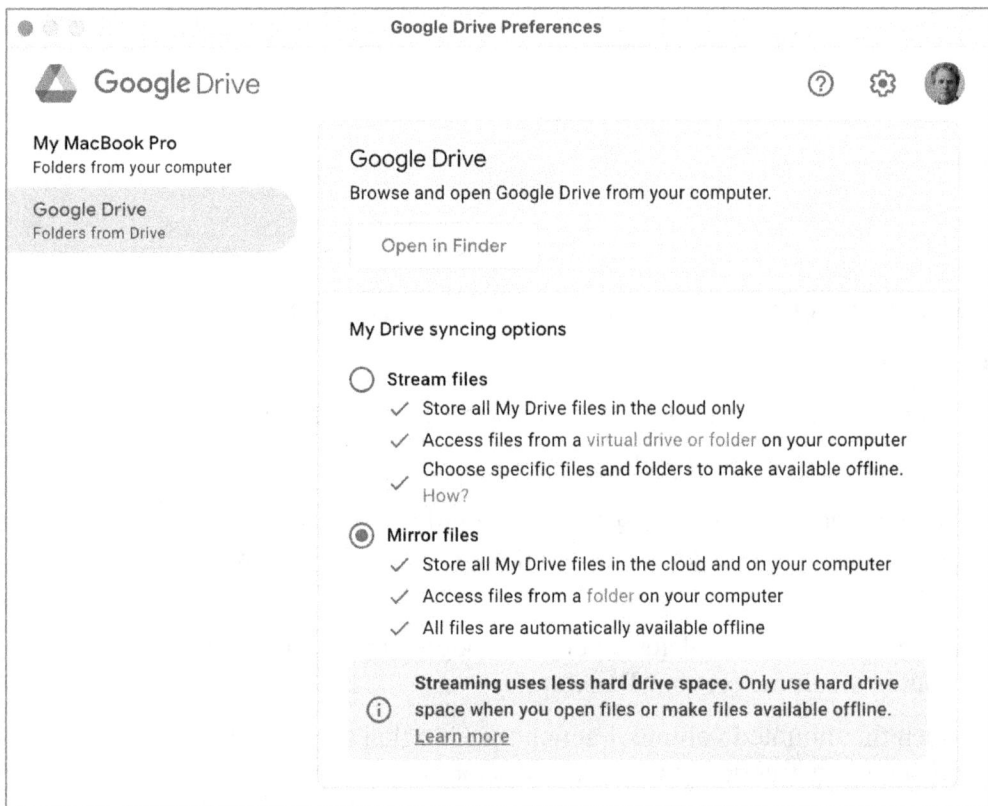

15. There are two synchronization methods–*Stream* and *Mirror.* Read what each offer, and then select how you wish to synchronize your data. My personal preference is to mirror. This gives me access to all my data even if I don't have an internet connection, and as all data resides on both my computer and

cloud, I can use a local backup such as Time Machine as my secondary backup.

16. When done, tap *Save* button.

17. From the Google Drive Menu bar, tap *Gear* icon > *Preferences.*

18. From the sidebar, tap *Google Drive* > *Gear* icon. The *Google Drive Preferences* window appears. Configure to your taste, then tap *Done* button:

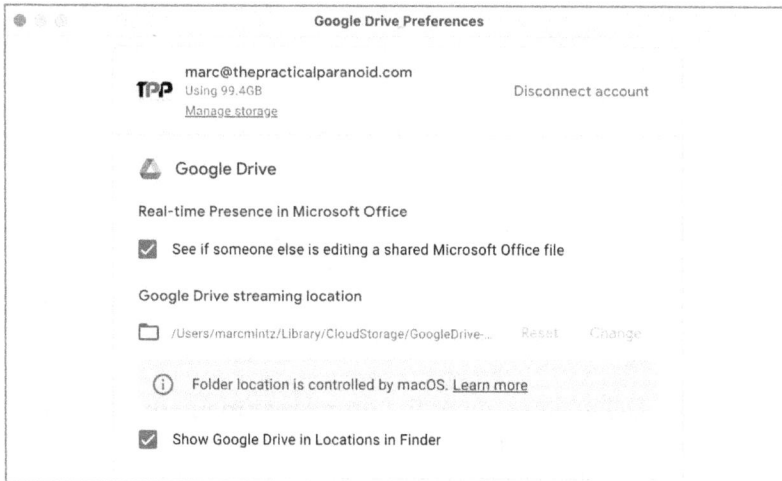

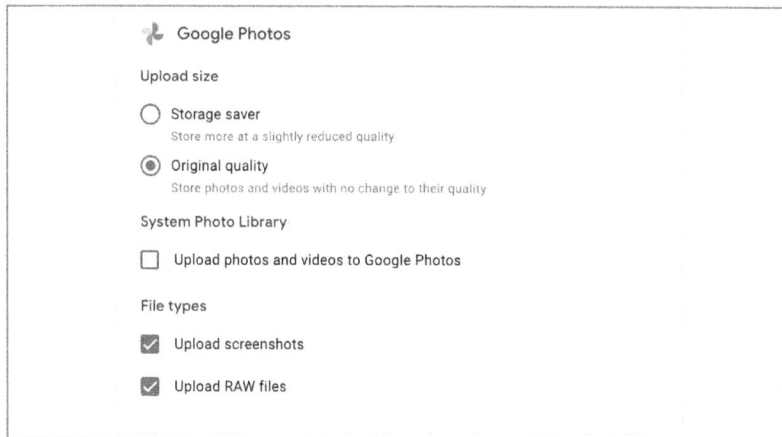

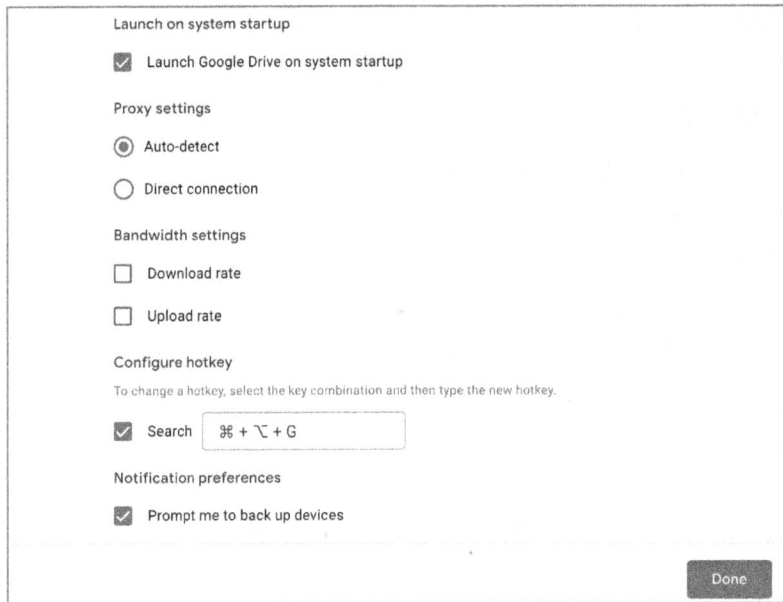

Monitor Google Drive Progress

19. From the *Google Drive* menu icon, the dropdown menu displays the most recent synchronized files, and a status report at the bottom.

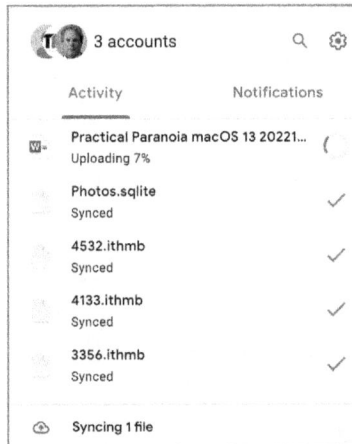

Accessing Your Data from Cloud-Based Google Drive

You can always access all your data from any device with a browser and internet connection.

20. Open a web browser to *https://drive.google.com.*

21. Log in with your Google email and password.

22. All your files and folders selected for synchronization on your device are found here.

2.1.16 [Optional] Assignment: Integrity Test Google Backup

In this assignment, you verify the integrity of your Google Backup.

1. Open *Google Drive.app.*

2. Tap the *Notifications* tab.

3. Verify there are no messages regarding synchronization errors.

4. Close *Drive.*

2.1.17 Assignment: Data Recovery from Google

Unlike with Android, there is no "data recovery" when backing up macOS to Google. Should you acquire a new device, simply installing the Google Drive app, and then authenticating in it, will give you access to all your data.

2.2 Back Up Your Cloud Data

Almost everyone uses the cloud. In the case of Chromebook users, virtually everything on their device is actually located on a Google server. Same for many Microsoft 365 users where their data may be located on Microsoft servers. MacOS users often have their entire Desktop and Documents folder contents stored on Apple servers.

But what happens when someone accidentally or maliciously deletes data from the cloud server? What happens when malware corrupts some (or all!) your data, and that data is either synchronized with a server or is not on the local drive but on the server?

Disaster is what happens.

The solution (before the disaster, not after) is to have a cloud-based backup of your cloud-based data!

There are many solutions available, with one sure to meet your needs. All will offer automated daily backup of email, contacts, calendars, and documents.

The services we have used and recommend include: AFI[16], Backupify[17], Cloudally[18], G-Suite Backup[19], Spanning[20], and SysCloud[21].

2.3 [Windows] Recovery Drive

2.3.1 [Windows] Assignment: Create A Recovery Drive

2.4 Data Loss Lessons Learned

- ☐ Data loss is a fact of life with many sources of loss including theft, power surges and sags, sabotage, fire, water damage, entropy, malware, physical shock, static electricity, criminal, and terrorist activities.

- ☐ Best practice calls for at least two backups, at least one onsite and at least one offsite.

- ☐ Internet-based data backup only backs up your data, not applications, OS, or settings.

- ☐ Internet-based data backup can take days or weeks to both upload and download data.

---

[16] *https://afi.ai*
[17] *https://backupify.com*
[18] *https://cloudally.com*
[19] *https://www.nucleustechnologies.com/g-suite-backup/*
[20] *https://spanning.com*
[21] *https://www.syscloud.com/*

☐ Storage devices should be formatted prior to first use to ensure they are in the proper format for the intended OS, and to verify the device is healthy.

☐ APFS disk format is best for current macOS with Fusion or SSD drive.

☐ OS X Extended (HFS+) disk format is best for macOS 10.14 and above with a HDD, and macOS 10.13 and earlier with a HDD, Fusion, or SSD.

☐ Time Machine is backup software included with macOS, which can create encrypted, non-bootable copies of all attached drives.

☐ iCloud can be used to automatically synchronize with your Desktop and Documents folders.

☐ Backups should be integrity tested monthly.

☐ Google Drive is a free backup service and application, limited to 15 GB data storage for all free Google accounts, and either 1 TB or unlimited data storage for all paid Google accounts.

☐ Every Apple ID comes with 5 GB of free iCloud storage space. This can be used with the free built-in iCloud backup software to keep the Desktop and Documents folders synchronized with iCloud.

3 Passwords

For a people who are free, and who mean to remain so, a well-organized and armed militia is their best security.

–Thomas Jefferson[1]

Knowledge, and the willingness to act upon it, is our greatest defense.

–Marc L. Mintz[2]

What You Will Learn in This Chapter

- Create a strong password
- Use the Keychain
- View an existing Keychain record
- Create challenge questions
- Store challenge Q&A in Keychain
- Access secure data from Keychain
- Harden the Keychain
- Synchronize Keychain across macOS and iOS devices
- Use Bitwarden to save website credentials
- Create password policies
- Perform 2-factor authentication

---

[1] *https://en.wikipedia.org/wiki/Thomas_Jefferson*
[2] *https://thepracticalparanoid.com*

What You Will Need in This Chapter

- [Optional] $10 for Bitwarden yearly subscription.

3.1   The Great Awakening

In June 2013, documents of NSA origin were leaked to The Guardian newspaper[3]. The documents provided evidence that the NSA was both legally and illegally spying on United States citizens' cell phone, email, and web usage. These documents, while causing gasps of outrage and shock by the public, revealed little that those of us in the IT field already did not know/suspect for decades: every aspect of our digital lives is subject to eavesdropping.

The more cynical amongst us go even further, stating that *everything* we do on our computers *is* recorded and subject to government scrutiny.

But few of us have anything real to fear from our government. Where the real problems with digital data theft come from are local kids hijacking networks, professional cyber-criminals who have fully automated the process of scanning networks for valuable information, competitors/enemies and malware that finds its way into our systems from criminals, foreign governments, and our own government.

The first step to securing our data is to secure our computers and mobile devices. Remember, we are not in Kansas anymore.

3.2   Strong Passwords

We all know we need passwords. But do you know that *every* password can be broken? Start by trying *a*. If that does not work, try *b*. Then *c*. Eventually, the correct string of characters gets you into the system. It is only a matter of time. Most attackers have more time than you do.

---

[3] *https://en.wikipedia.org/wiki/NSA_warrantless_surveillance_controversy*

Way back in your great-great-great grandfather's day, the only way to break into a personal computer was by manually attempting to guess the password. Given that manual attempts could proceed at one attempt per second, an 8-character password became the standard. With a typical character set of 26 (a–z) this created a possibility of 26^8 or over 100 billion combinations. The thought that anyone could ever break such a password was ridiculous, so your ancestors became complacent.

A while back the first hacker wrote password-breaking software. Assuming it may have taken eight CPU cycles to process a single attack event, on an old computer with a blazing 16 KHz CPU that equated to 2,000 attempts per second. This meant that a password could be broken in less than 2 years.

All of this is funny when you consider that research has shown most passwords can be guessed. Most computer users are unaware that what they thought was an obscure and impossible-to-break password could be cracked in minutes.

Yikes. IT directors took notice.

Down came the edict from the IT Director that we *must* create *obscure* passwords: strings that include upper and lower case, numeric, and symbol characters. But in many cases, this was a step backward. Since a computer user could not remember their password was 8@dC%Z#2, the user often would manually record the password. That urban legend of leaving a password on a sticky note under the keyboard? I have seen it myself more than a hundred times.

In 2003, Bill Burr came up with a set of password security guidelines for NIST. These included requirements to include a number, a capital letter, and a symbol like an asterisk. Unfortunately, that guidance was wrong. If you have an 8-character password, but one letter must be a capital, then that one letter has only 26 possible values instead of 72 (26 lower-case letters, 26 upper-case, 0-9 and 10 special characters). You haven't made your password safer – you've reduced the *key space* and made it easier to crack! Require a number, and that character only has 10 possible values – making your password easier to crack still. Require a special character (assuming the 10 most common such as @) and your password is yet easier to crack. In 2017, Mr. Burr issued a massive apology[4] because his advice had made passwords weaker, not stronger.

---

[4] *https://www.wsj.com/articles/the-man-who-wrote-those-password-rules-has-a-new-tip-n3v-r-m1-d-1502124118*

Present day: A current personal computer with freely available password-cracking software can make over 10 billion password attempts per second. Create an army of infected computers called a botnet to do your dirty work[5] and you can achieve over a hundred trillion attempts per second, unless your system locks out the user after x number of failed logon attempts. That's critical, but while local applications and network logons are likely to use a preset lockout rate (say, three unsuccessful attempts), many web sites and online applications do not.

What does this army of enemy computers mean for you? The typical password using upper and lower case, number, and symbol now can be cracked with the right tools in under two minutes. If the hacker uses just a single computer to do the break in, make that a whole week. Don't believe it? Look at the *Bitwarden Password Strength Testing Tool*[6].

It's true that if we use longer passwords, we can make it too time consuming to break into our system, so bad guys may move on to other targets. But you say it is tough enough to remember *eight* characters, impossible to remember more?

That's only true if we keep doing things as done before. Since attacks now are done by automated software, password hacking is only an issue of length of password, not complexity. So, forget memorizing meaningless combinations of digits and use a passphrase that is easy to remember, such as, "Rocky has brown eyes" which at 100 trillion attempts per second could take over 1,000,000,000,000,000 centuries to break (provided Rocky is not the name of your beloved pet and thus more guessable).

How long should you make your password, or rather, passphrase? My recommendation to clients is a minimum of 15, in an easy-to-enter phrase. As of this writing, Apple[7] and Microsoft[8] recommend a minimum of 8 characters, while Google[9] recommends 12. US-CERT[10, 11] recommends at least 15 for

---

[5] *http://en.wikipedia.org/wiki/Botnet*

[6] *https://bitwarden.com/password-strength/*

[7] *https://support.apple.com/en-us/HT201303*

[8] *https://www.microsoft.com/en-us/research/wp-content/uploads/2016/06/Microsoft_Password_Guidance-1.pdf*

[9] *https://support.google.com/a/answer/33386?hl=en*

[10] *https://security.web.cern.ch/security/recommendations/en/passwords.shtml*

[11] *https://www.us-cert.gov/ncas/alerts/TA11-200A*

administrative accounts, at least 8 for non-administrators. Cisco recommends[12] at least 8.

In addition to password length, it is critical to use a variety of passwords. In this way, should a bad actor gain access to your Facebook password, that password cannot be used to access your bank account. There's the famous story of Mat Honan[13], who had worked for Gizmodo and WIRED Magazine (he should have known better, but this was back in 2012 when dinosaurs still roamed), whose iCloud account was hacked. That's bad enough, but Mat had used the same password on Gmail, Twitter, and other accounts – all of which also were hacked in less than an hour.

Yes, soon you have a drawer full of passwords for all your different accounts, email, social networks, financial institutions, etc. How to keep all passwords organized and easily accessed amongst all your various computers and devices? More on that later in this chapter.

Microsoft and Apple are encouraging users to move away from passwords and move toward fingerprint, facial recognition, and other (possibly more secure) options. It would be prudent to mention that biometrics are only more secure against remote attacks. In person, someone could get control of your finger, they could use it to access the machine. Same goes for your face. May seem unlikely but there are cases before the courts of law enforcement officers doing just that and more to get biometric access to systems without a warrant. As a bonus, the courts have been inconsistent[14] when ruling on the legality and constitutionality of said cases.

Our current recommendation is to NOT use biometrics to log in or wake any device, but once logged in, it is acceptable to use biometrics for purchases, and as an alternative to entering passwords.

---

[12] *http://www.cisco.com/c/en/us/td/docs/ios-xml/ios/sec_usr_aaa/configuration/15-sy/sec-usr-aaa-15-sy-book/sec-aaa-comm-criteria-pwd.html*
[13] *https://www.wired.com/2012/08/apple-amazon-mat-honan-hacking/*
[14] *https://9to5mac.com/2020/08/12/police-demand-you-unlock/*

2023 Apple Password Recommendations[15]

- Minimum 8-character length

- Include at least one number

- Include both upper and lowercase letters

- For a stronger password, add additional characters

2023 Google Password Recommendations[16]

- Use a different password for every site

- Minimum 12 characters long

2023 Microsoft Password Recommendations[17]

- Minimum 8-character length

- Contain characters from at least three of the following: Uppercase letters, lowercase letters, base 10 digits, special characters, any Unicode character that is categorized as an alphabetic character but is not uppercase or lowercase

- Must not contain the user's account name value

- Must not contain the user's display name

- Eliminate character-composition requirements

- Eliminate mandatory periodic password resets for user accounts

- Ban common passwords, to keep the most vulnerable passwords out of your system

- Educate users not to re-use a password for non-work-related purposes

- Use multi-factor (2-factor) authentication

---

[15] *https://support.apple.com/en-us/HT201303*

[16] *https://support.google.com/accounts/answer/32040?hl=en#zippy=%2Cmake-your-password-unique%2Cmake-your-password-longer-more-memorable*

[17] *https://docs.microsoft.com/en-us/windows/security/threat-protection/security-policy-settings/password-must-meet-complexity-requirements*

2023 US-CERT Password Recommendations[18, 19]

- Private and known only by one person
- Not stored in clear text in any file or program, or on paper
- Easily remembered
- Minimum 15-character length for administrators, minimum 8-character length for non-administrators
- A mixture of at least 3 of the following: upper case, lower case, digits, and symbols
- Not listed in a dictionary of any major language
- Not guessable by any program in a reasonable time frame

3.2.1 Assignment: Create a Strong Password

In this assignment, you test the strength of a password for your device or website.

Test password strength

1. Think up a password for yourself that consists of at least 15 easy-to-enter characters and meets the strength/complexity required by your organization.

2. Open a web browser to *https://bitwarden.com/password-strength/*. Scroll down to the *Evaluate Your Password* entry field. Then enter your password in the field.

---

[18] *https://www.us-cert.gov/ncas/tips/ST04-002*
[19] *https://www.us-cert.gov/ncas/alerts/TA11-200A*

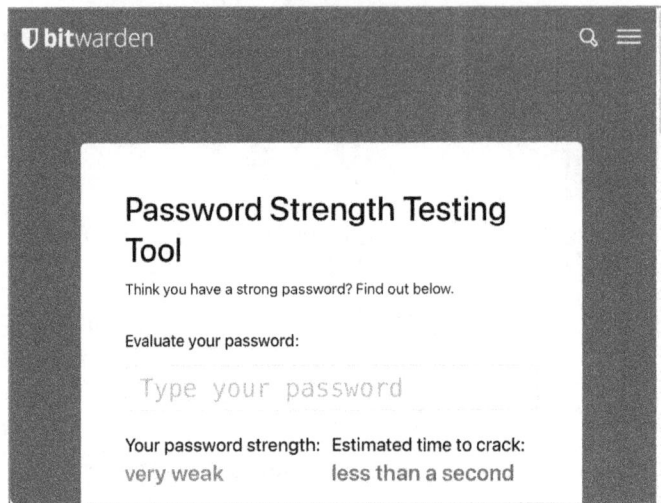

3. Scroll down to the *Your password strength* and *Estimated time to crack*.

4. Repeat with several more of your passwords.

5. If you find passwords that do not meet your security needs, change them now.

6. Record your new password.

7. Exit the browser.

Change your old password to the strong password

8. If the password is for a website, visit it now and change the password.

3.2.2   Assignment: Create a New Login Password

In this assignment, you change your device login password to a stronger password.

1. Tap on *Apple* menu > *System Settings* > *Login Password*.

2. Select the *Change...* button.

3. Enter your *Old password*, your *New password*, *Verify*. Then select *Change Password* button.

4. Quit *System Settings*.

Your new, strong password now is in effect.

3.2.3 [iOS and Android] Assignment: Erase Data After 10 Failed Passcode Attempts

3.2.4 [Windows] Assignment: Sign-In Options

3.3 Password Manager

In our grandparent's day, life was so much simpler. To give an example: My grandfather always had four keys in his pocket: one for home, one for the car, and the other two he could never remember what for.

In today's world, the realm of keys has expanded into the digital world. You now have keys or passwords for logging onto your computer, phone, tablet, email, many of the websites you visit, Wi-Fi access points, servers, frequent flyer account, etc. In my case, I have 857 passwords in use. I know because they are all neatly stored in a database so that I do not have to remember them.

Unfortunately for most of us, our "keys" are not very well organized, so when we need to access our mail from another computer, or order a book on Amazon, we are stuck.

By default, your Mac stores most usernames and passwords that you use to access Wi-Fi networks, servers, other computers, and websites. The exceptions are websites programmed to not have credentials saved, such as financial institutions.

The built-in tools that store this information automatically also can be used to manually store any text-based data. This includes credit card information, software serial numbers, challenge Q&A, offshore banking information, etc.

macOS has two locations to store keys:

- Web browser, which stores only credentials for websites visited with that browser.

- Keychain database, which stores username, password, and URL for websites that request authentication, Wi-Fi networks, servers, other computers you access, email accounts, and encrypted drives.

  o Located at *~/Library/Keychain*

  o Opened with the *Keychain Access* utility, located in */Utilities*

- Note: New as of iOS 15 and macOS 12 is the ability to share Keychain passwords with Windows users with iCloud for Windows, and iCloud Passwords browser extension for Chrome and Edge.

Keychain is what interests us here.

Let's take the case of visiting a website that requires a username and password, connecting to another computer or server, or performing some action that triggers an authentication request. The following are the steps as they typically occur:

1. A prompt appears requesting a username and password.

- Typical default authentication window for a server:

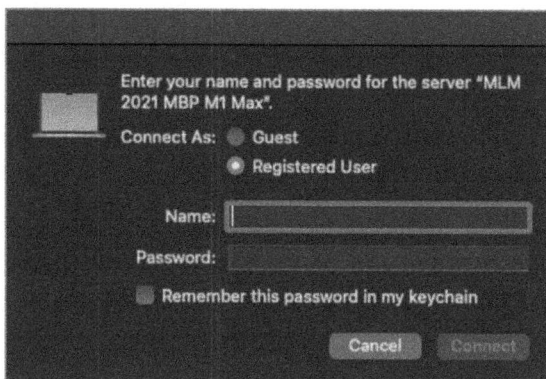

- Typical authentication window for a website:

2. Enter your username and password.

   - If there is a checkbox to *Remember this password in my Keychain,* enable it to copy your credentials into the Keychain.

   - The browser may offer to save the password for you. In this case, the credentials are stored in the browser's database.

3. The website takes you to the appropriate secured page, or the other computer mounts a drive on your Mac.

Behind the curtain, your Mac has copied your username and password into the Keychain database, named *login.keychain-db*, located in the Home *Library/Keychains* folder. The database is military grade AES 256 encrypted (using your login password), safe from prying eyes.

- Note: This Home Library folder is normally invisible. It can be accessed by holding down the *option* key while selecting the Finder *Go* > menu. The next time you visit this same website or server, the steps change to this:

1. You surf to the website or select a server to access.

2. A prompt appears requesting a username and password.

3. A query is made of the browser database and the Keychain database based on the URL of the site or the name of the server.

4. If either the browser or the Keychain database has stored the username and password associated with the URL or server, the credentials are automatically copied/pasted into the username and password fields.

5. Select *Enter*.

6. The website takes you to the appropriate secured page or the server share point mounts.

Note that you did not need to know your credentials. Keychain did it all for you.

Passwords System Setting

Introduced with macOS 12 is the *Passwords System Setting.* This offers a much faster and easier interface to work with your passwords.

In addition, Windows devices with the new *iCloud Passwords* app (included with iCloud for Windows) can access and use passwords stored in iCloud.

With the new *Passwords Extension for Edge,* passwords stored in iCloud are available for autofill in Edge.

3.3.1   [Android] Assignment: Enable Autofill with Google

3.3.2   Assignment: View an Existing Keychain Record

For this assignment, let us assume you have forgotten the password to your Wi-Fi network. The Keychain database has stored the password. You just need to look for it.

In this assignment, you examine a record in the Keychain.

1. Launch *Keychain Access* (located in /Applications/Utilities/).

2. From the sidebar, in the *Keychains* field, select *iCloud* if you have enabled *Keychain* in *System Settings > Apple ID*), or *login* if you have not enabled this.

- *iCloud* keychain holds the credentials that are synchronized with all your other devices through iCloud.

- *Login* keychain holds the credentials that are local to only this device.

3. In the center, main area of the window, double tap on the target Wi-Fi network.

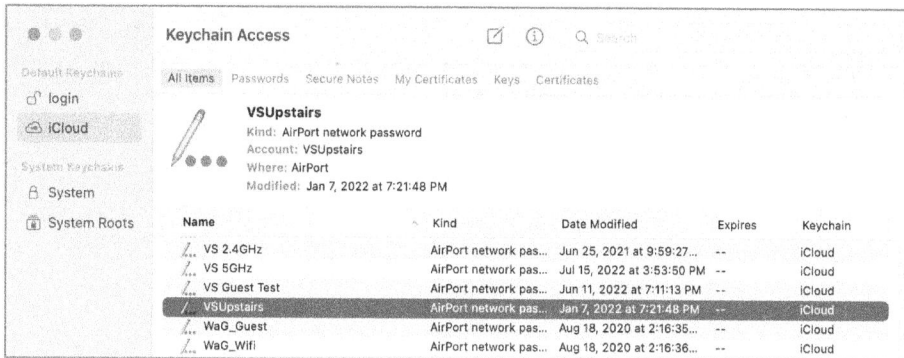

4. The records *Attributes* window opens.

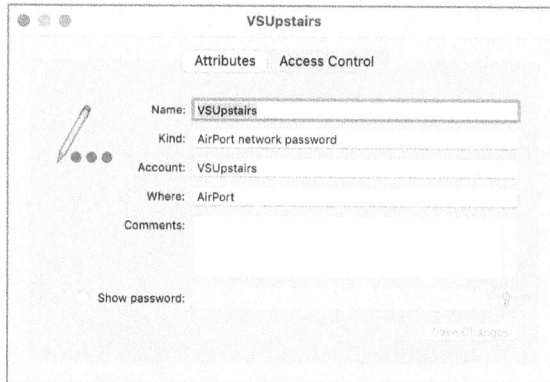

5. At the bottom of the *Attributes* window, enable the *Show Password* checkbox. This opens the authentication window.

6. Enter your Keychain password. This is the same as your user account password, then tap the *Allow* button. This authorizes Keychain to show you the password.

7. The *Show Password* field displays the needed password.

8. Quit *Keychain Access*.

3.3.3   Assignment: Keychain Auto-Entry

In this assignment, you have Keychain auto-enter username and password for a website.

- Prerequisite: Website credentials stored in Keychain.

1. Open Safari on your device, then go to a website that requires credentials (username and password) and has them stored in Keychain. For this example, I go to *https://appleid.apple.com.*

2. The authentication window opens.

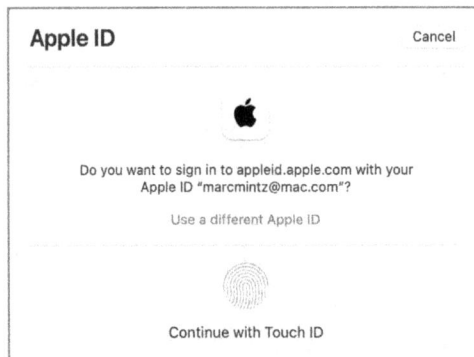

3. If your computer has TouchID, tap your fingerprint sensor to auto-enter your credentials. *Use a different Apple ID* to get a prompt to enter your Apple ID.

4. The username is auto filled.

5. If requested, repeat for your password.

6. Your credentials are entered without you having to even know them!

3.3.4   [Optional] Assignment: Harden Keychain Security

By default, the Keychain is unlocked when the user logs into their computer and remains unlocked until the user logs out.

In some high-security environments it may be desirable to increase the security of Keychain. There are two options available: *Lock after ____ minutes of inactivity,* and *Lock when sleeping.* Both options require the user to enter their computer account password to unlock Keychain after either X minutes of computer inactivity or after waking from sleep.

In this assignment you will see how to harden Keychain security.

1. Open *Keychain Utility.app.*

2. From the sidebar, select *login.*

3. Select *Edit* menu > *Change Settings for Keychain "login".* The *Settings* window opens.

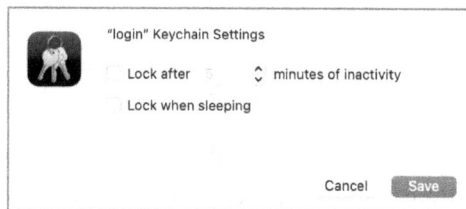

4. Configure to your taste, select the *Save* button, then *Quit* Keychain Access Utility.

3.3.5   Assignment: Use Passwords System Setting to View Passwords

In this assignment, you explore the *Passwords* System Setting.

1. Open *Apple* menu > *System Settings* > *Passwords.*

2. Enter your login password. The *Passwords* System Preference window opens.

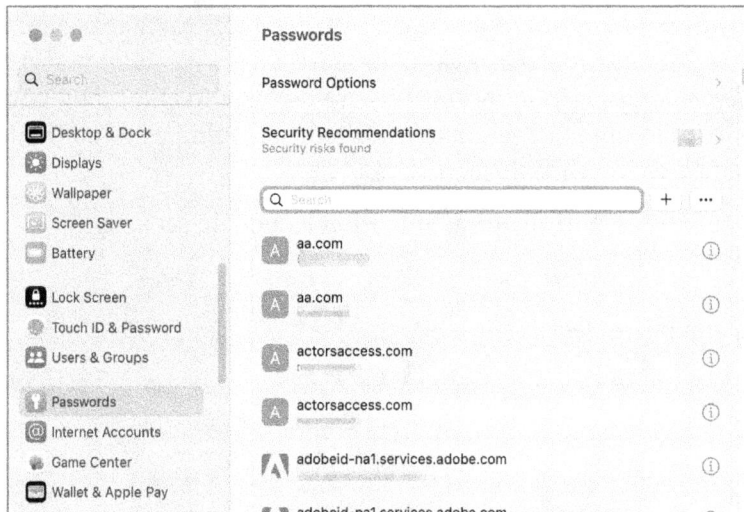

3. The main area displays your accounts with stored credentials.

- Note: Passwords that are used in more than one site, easily guessed, or compromised can be found by selecting *Security Recommendations*.

4. To view or edit a password:

  a. Double tap the target site.

  b. To view or edit the password, tap the *Edit* button.

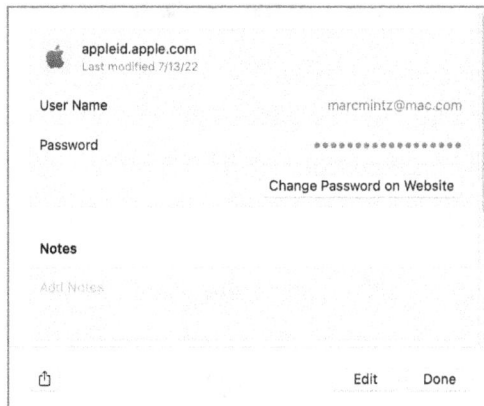

  c. When done, tap the *Done* or *Save* button.

3.3.6 Assignment: View Credit Card Information

The Keychain also stores credit cards you used in Safari. In this assignment, we view any credit card information stored on your device.

1. Open *Safari.*

2. Tap *Safari* menu > *Settings* > *Autofill* > *Credit Cards* > *Edit* button.

3. At the authentication prompt, enter your device password.

4. Any credit cards that have been saved now appear.

5. Exit out of *Safari Settings*.

Congratulations! You now know more about the keychain than you ever wanted to know.

3.4   Challenge Questions

A Challenge Question is a way for websites to authenticate who you claim to be when you contact support because of a lost or compromised password.

For example, when registering at a website you may see an entry for: *Question – Where did your mother and father meet?*

The problem with this strategy is that most answers are easily discovered with an Internet search of your personal information, or a bit of social engineering.

The solution is to give bogus answers. For example, my answer to the question; *Where did your mother and father meet?* might be: *1954 Plymouth back seat.* It would not be possible for a hacker to discover this answer, as it is completely bogus. My mother tells me it was a 1952 Dodge.

Unless you are a savant, there is no way to remember the answers to all challenge questions. But there is no need to remember. We already have a built-in utility that is highly secure and designed to hold secrets such as passwords: Keychain Access!

Although Keychain can automatically record and autofill usernames and passwords, to further protect you it requires manually entering other data such as challenge Q&A.

3.4.1   Assignment: Store Challenge Q&A in the Keychain

In this assignment, you manually store the challenge Q&A for a pretend website: myteddybear.com.

1. Open *Keychain Access.app*, located in */Applications/Utilities*.

2. From the side bar, tap *iCloud*.

    o   Note: By creating the secure note in the iCloud section, the note will synchronize to your iCloud Keychain. If created within the default *Login* section, the secure note will be accessible only to this device.

3. Select the Keychain Access *File* menu > *New Secure Note Item*.

4. The Keychain *Item Name* window appears.

5. In the *Keychain Item Name* field, enter: *myteddybear.com Q&A*.

6. In the Note field, enter:
    Q: Where did your parents meet? A: I don't know
    Q: What is the name of your first pet? A: Swims with fishes
    Q: What is the name of your high school? A: Who needs an education

7. Select the *Add* button.

8. Find your new Secure Note within all your other Keychain items.

9. Double tap on the Secure Note to open it.

10. Enable the *View Note* check box.

11. In the authentication window, enter your system login password, and then tap *OK*. The secure note opens to display your Q&A information.

12. Quit Keychain Access.

Your challenge questions and answers now are securely stored.

3.5 Synchronize Keychain Across iOS, macOS, and Windows Devices

Perhaps like me, you may need to access most of these passwords anywhere, anytime. It also is a huge help to have the credentials created on your macOS computer available to your iOS device, and vice versa.

If you have macOS X 10.9 or higher or macOS and iOS 7 or higher, to a large degree Apple has you handled. With the most recent incarnations of both operating systems, Apple has added *Keychain* to the iCloud synchronization scheme. This allows your Keychain database to be synchronized between all your Apple computers, iPhones, and iPads. New with iOS 15 and macOS 12 is Keychain synchronization with *iCloud for Windows* and Google Chrome and Microsoft Edge browser extensions.

3.5.1 Assignment: Activate iCloud Keychain Synchronization

In this assignment, you enable iCloud Keychain synchronization between all your iOS and macOS devices.

1. Open the *Apple* menu > *System Settings* > *Apple ID*.

2. Enable the *Keychain* checkbox. The *Enter your Apple ID password to setup iCloud Keychain* dialog box appears.

3. Enter your *Apple ID password*, then select the *OK* button.

4. If you have previously created a 2-step verification (also known as *2-factor authentication* or *multi-factor authentication*) for your Apple ID, the *Keychain Setup* dialog box opens. Select the *Request Approval* button.

5. A request is sent to the other devices currently approved on your account to approve this device. Enter your Apple ID password, then tap *Allow*.

6. Go back to *System Settings* and notice that the *Keychain* now is enabled.

7. *Quit* System Settings.

Your Keychain on this computer now synchronizes automatically with your iCloud account, and therefore with all other macOS, and iOS devices synchronizing on the same account.

3.6   Password Management with Bitwarden

A great solution to the need for credential, challenge question, and Time-based One-Time Password (TOTP) management is the open-source utility Bitwarden[20]. Bitwarden features the following:

- Secure password sharing

- Security audit and compliance

- 24/7 support

- Cross-platform operation

- Cloud-based or self-hosted options

- Fully functional free version, as well as for-fee version

- Two-Factor Authentication (Bitwarden calls this Time-Based One-Time Password, or TOTP) (requires for-fee version)

3.6.1   Assignment: Install and Configure Bitwarden

In this assignment, you install the free version of Bitwarden.

- Note: If you use only Apple devices and browser, this assignment is optional, as the major advantage of Bitwarden is the ability to work cross platform and with browsers other than Safari.

Download and Install

1. Open *App Store,* search for then download and install Bitwarden.

2. Open *Bitwarden*, then tap *Create Account.*

3. Enter the requested information to create an account, then tap *Submit.* This returns you to the login screen.

Configure

4. Log into Bitwarden.

---

[20] *https://bitwarden.com*

5. Select *Bitwarden* menu > *Settings.* Configure to your taste. My recommendation is shown below:

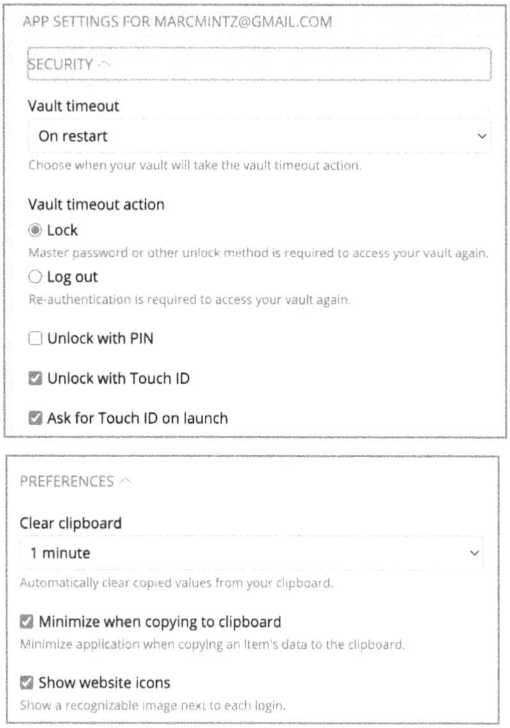

APP SETTINGS FOR MARCMINTZ@GMAIL.COM

SECURITY ^

Vault timeout

On restart   ⌄

Choose when your vault will take the vault timeout action.

Vault timeout action

◉ Lock

Master password or other unlock method is required to access your vault again.

○ Log out

Re-authentication is required to access your vault again.

☐ Unlock with PIN

☑ Unlock with Touch ID

☑ Ask for Touch ID on launch

PREFERENCES ^

Clear clipboard

1 minute   ⌄

Automatically clear copied values from your clipboard.

☑ Minimize when copying to clipboard

Minimize application when copying an item's data to the clipboard.

☑ Show website icons

Show a recognizable image next to each login.

APP SETTINGS (ALL ACCOUNTS) ∧

☑ Show menu bar icon
Always show an icon in the menu bar.

☐ Minimize to menu bar
When minimizing the window, show an icon in the menu bar instead.

☐ Close to menu bar
When closing the window, show an icon in the menu bar instead.

☐ Start to menu bar
When the application is first started, only show an icon in the menu bar.

☑ Start automatically on login
Start the Bitwarden desktop application automatically on login.

☑ Always show in the Dock
Show the Bitwarden icon in the Dock even when minimized to the menu bar.

☑ Allow browser integration
Used for biometrics in browser.

☑ Allow DuckDuckGo browser integration
Use your Bitwarden vault when browsing with DuckDuckGo.

☑ Require verification for browser integration
Add an additional layer of security by requiring fingerprint phrase confirmation when establishing a link between your desktop and browser. This requires user action and verification each time a connection is created.

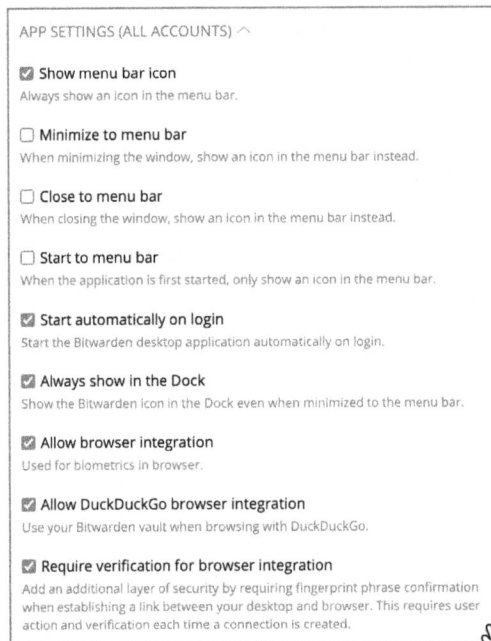

6. When complete, tap *Close*.

Enable Two-Step Login

As the keys to your treasure are stored in this database, not only is a strong Bitwarden password important, but so is having Two-Step Login (2-Factor Authentication) enabled.

7. Open a browser to *https://bitwarden.com*.

8. Tap *Log In,* enter your login credentials then tap *OK*.

9. From the Bitwarden toolbar, tap *Settings > Two-Step Login*.

10. Select your preferred method to get a verification code. In this assignment you use *Email*.

11. At the prompt, enter your Bitwarden email address, then tap *Send Email*.

12. Open your email to find the verification email.

13. Copy the verification code from your email, paste it into the verification field, then tap *Enable*.

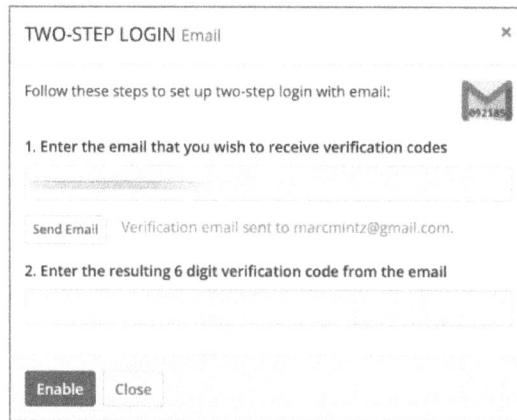

14. At the dialog box stating that *This Two-Step Login provider is enabled on your account,* tap *Close.*

15. In the Bitwarden *Two-Step Login* webpage, tap *View Recovery Code.*

16. Enter your Bitwarden password, then tap *Continue.*

17. In the *TWO-STEP LOGIN Recovery code* window, copy then securely store your recovery code. When done, tap *Close.*

   - Note: In the event you no longer have access to your Bitwarden Authenticator, this is the only way you can regain control over your account.

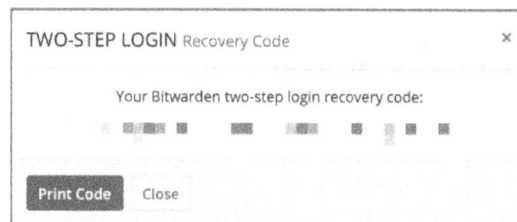

Install and Configure Safari Extensions

To automatically work with your browser, a Bitwarden Browser Extension must be installed for each browser.

18. If Safari is currently open, Quit Safari.

19. Open Safari.

20. If the *Bitwarden* login window appears, enter your Bitwarden credentials, then tap *Log In.*

21. Open *Safari > Preferences > Extensions* tab.

22. Enable the *Bitwarden* extension.

23. At the verification prompt, tap *Turn On.*

24. Tap the *Always allow on every website* button.

25. At the verification window, tap *Always allow on every website* button.

26. Close *Preferences.*

27. In the Safari toolbar, tap the Bitwarden icon.

28. Tap *Login*, then enter your Bitwarden credentials.

29. Tap the Safari Bitwarden toolbar icon > *Settings.*

30. Tap *Sync > Sync Vault Now* to synchronize passwords.

31. Configure *Vault Timeout* to *On Browser Restart,* and *Vault Timeout Action* to *Lock.*

32. Scroll down to tap *Options.* Configure to your taste. My recommendation is to leave all settings at the default, except to enable *Enable Autofill On Page Load.*

33. When done, tap outside of the Bitwarden window to save and close it.

Install and Configure Brave Extension

- Note: Brave is a Chromium-based open-source browser, designed with security and privacy as the top priority. If you do not yet have Brave installed, you can do so by visiting *https://brave.com.*

34. Open Brave.

35. Browse to *https://bitwarden.com.*

36. Tap *Download.*

37. Scroll to the *Web Browser* area.

38. Tap on the desired browser extension. In this assignment, Brave.

39. Tap *Add to Brave* button.

40. In the Brave alert, enable *Allow this extension to run in Private Windows,* then tap *Okay, Got it.*

41. In the Brave toolbar, tap the *Bitwarden* icon > *Log In.*

42. Enter your Bitwarden credentials, then tap *Return* key to login.

43. In the Brave toolbar, tap the *Bitwarden* icon *Settings* > *Sync* > *Sync Vault Now.* When done, tap outside of Bitwarden to close.

44. Configure your Bitwarden *Options* same as your configured them in Safari.

3.6.2 Assignment: Enter New Credentials in Bitwarden

Credentials may be entered either manually or automatically into Bitwarden.

In this assignment you manually enter new credentials into Bitwarden.

• Prerequisite: Bitwarden account, Bitwarden installed on your device, and Bitwarden browser extension configured and logged in.

Manual Entry

1. Open a browser.

2. From the browser toolbar, select the Bitwarden icon > +.

3. In the *Add Item* window, Enter the *URL* for the site. In this assignment, use *www.facebook.com.*

4. Enter your desired *Username* and *Password.*

5. Scroll down to tap on *New URI.*

6. Enter the full URL for the site. For this example, enter `https://www.facebook.com.`

7. Tap *Save.*

8. In the browser, visit *https://www.facebook.com*

9. Tap *Log In.* Your credentials autofill, and you are now logged in.

10. Log out of Facebook.

Automatic Entry

11. Visit a website that is not stored in Bitwarden.

12. Enter your site login credentials, then tap *OK*.

13. Bitwarden pops up a dialog asking if you want to save these credentials. Tap *Yes*.

3.7   Password Policies

A password policy is a set of rules to help users create and use passwords. You have seen password policies in use when creating a password for your online banking or shopping and were alerted that your password needed to be longer or have a special character.

Within the government, military, financial, and healthcare environments, and the NIST 800-171 security protocols, setting *password policies* is a mandate. Although not a mandate for the home and general business computer, doing so makes a lot of sense.

In an IT environment which is controlled by a Microsoft Active Directory server, password policies can be enforced from the server. In environments without a server, password policies can be enforced using either the *Apple Configurator.app* or Terminal for command-line control.

Apple Configurator is available for free download to macOS computers from the Apple App Store. It allows the creation of *profiles*, which force behaviors over the Macintosh computer, iPhone, and iPad. If you are responsible for supporting more than just your own computer, it is best practice to ensure all computers are configured with password policies.

3.7.1   Assignment: Password Policies with Apple Configurator

In this assignment, you install Apple Configurator, then create a password policy saved as a profile.

1. Open *Apple* Menu > *App Store*.

2. In the App Store, search for *Apple Configurator 2*. Once found, download it.

3. Launch *Apple Configurator*.

4. Select *File* menu > *New Profile.* The new, untitled profile editor appears.

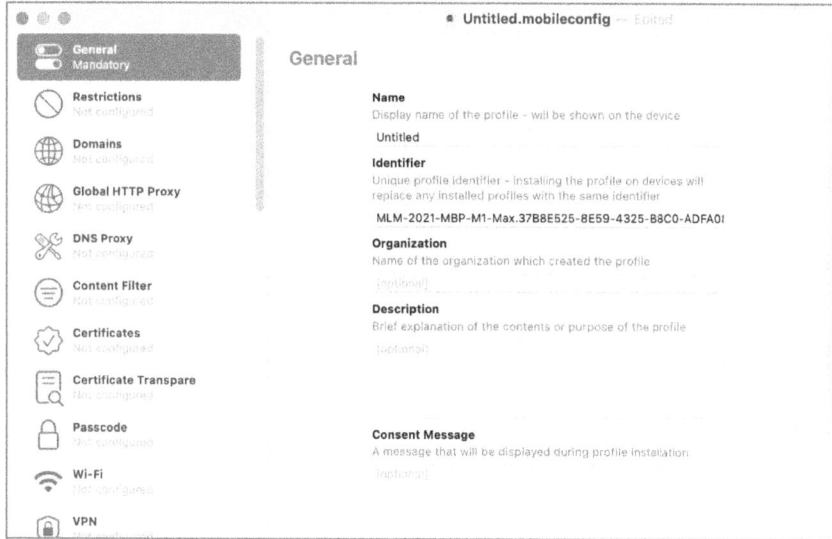

5. In the sidebar, select *General*. The *General* settings window appears.
    Configure as below:

    ● Name: *Password Policies*

    ● Identifier: *password policy*

    ● Security: *With Authorization*

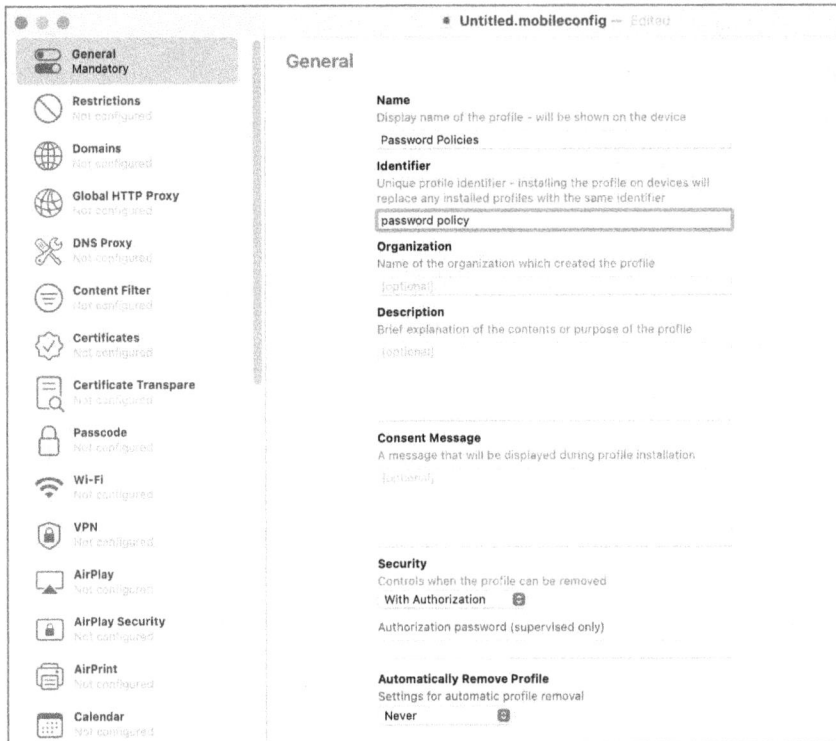

6. In the sidebar, select *Passcode*, then in the *Passcode* window, tap *Configure.*

7. For this assignment, configure as below:

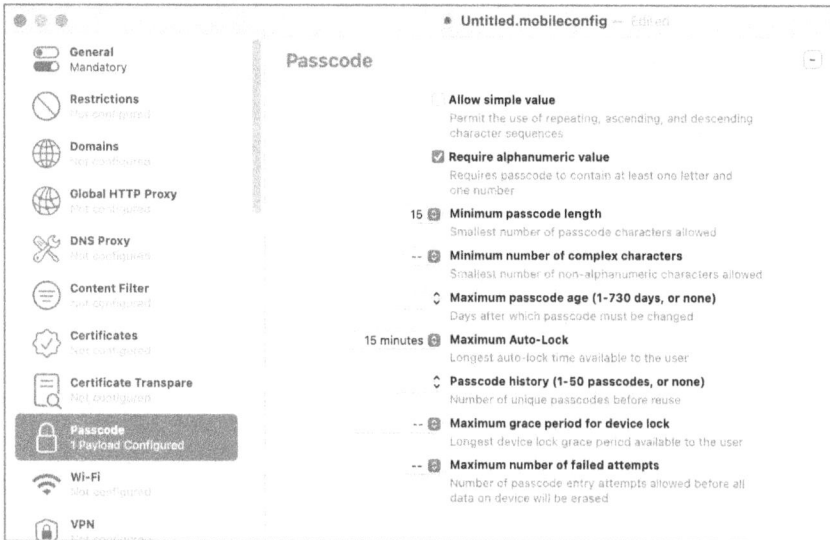

8. Select *File* menu > *Save*.

9. In the *Save As* window, name the profile *PasswordPolicies.mobileconfig*, then save to your desktop.

10. Quit Apple Configurator.

3.7.2   Assignment: Install Password Policies Profile

In this assignment, the password policy profile created in the previous assignment is installed on your computer.

● Prerequisite: Completion of the previous assignment.

Install the policy

1. Double tap the *PasswordPolicies.mobileconfig* file saved to your Desktop in the previous assignment.

2. If prompted to *Install Password Policies?*, tap *Continue*.

3. Tap this window when it appears: *Profile Installation – Review the profile in System Settings if you want to install it.*

> **Profile installation**
> Review the profile in System Settings if you
> want to install it.

4. Open *System Settings* > *Privacy & Security* > *Profiles*.

5. Double-click on the new *Password Policies* policy. A display of all policy functions will appear. Review the policy functions.

Are you sure you want to install this profile?	
Require Password	True
Allow Simple	False
Require Alphanumeric	True
Min Password Length	15
Auto-Lock (minutes)	15
Install...	Ignore Cancel

6. Tap the *Install* button to install the policy.

7. At the *Are you sure you want to install profile "Password Policies"?* prompt, Tap *Install*.

8. Enter an administrator's credentials, then tap *OK*.

9. The profile now is installed and active.

Test the password policy profile

10. Open *System Settings* > *Users & Groups*.

11. Select the currently logged-in account.

12. Select *Change Password*.

13. In the *Change Password* window:

   a. Enter your old password.

   b. Enter a new password that does not meet the password policy minimum requirements.

   c. Tap in the *Verify* field. An alert appears forcing minimum password policy requirements.

Remove the password policy profile

If you prefer not to have this profile active on your computer, you can remove it.

14. Tap *Cancel*.

15. In System Settings, select *Profiles*.

16. Select the *Password Policies* profile.

17. Select the – (remove) button.

18. At the *Are you sure you want to remove the locked profile* window, select *Remove*.

19. At the authentication prompt, enter an administrator's name and password, then select *OK*. The policy is deleted.

Install this password policy onto other Macs

20. If you want to install this password policy on other computers, copy the *PasswordPolicies.mobileconfig* file to those computers, and repeat the steps above.

3.8   2-Factor Authentication

Maintaining strong passwords that are unique to every site is the foundation for site and account security. But it is only the start.

Every password can be cracked. Many websites do not properly encrypt passwords. When such a site is itself hacked, millions of user account and password credentials are harvested in seconds.

The fix for these issues is the use of *2-Factor Authentication (2FA)*, sometimes referred to as *Multi-Factor Authentication* or *2-Step Authentication*. With 2FA active, not only are the account username and password required for access, but a second form is also required. This can take the form of:

- An email.

- SMS or text message of a code to your phone.

- o Note: This is no longer considered high security, as it is possible to intercept such text messages.
- 2FA random code generation performed by an app.
- 2FA random code generation performed by a hardware device (typically a USB stick).

There are dozens of 2FA apps available, with the *Google Authenticator* being the most popular. However, my recommendations go to *Bitwarden*[21] and *Passwords System Preference* (macOS 12 and higher). Their benefits include:

- Easy 2FA setup for any account
- High security, with Touch ID, Face ID, and PIN protection
- Cloud-based encrypted backup
- Use on multiple devices
- Bitwarden works cross platform with Android, Chrome OS, iOS, macOS, and Windows.
- Passwords works with macOS 12 and higher and iOS 15 and iPadOS 15 and higher.
- Note: Bitwarden is open-source and free, but to enable its Authenticator module (as well as a few other features) requires upgrading to a for-fee tier, ranging from $10/year to $40/year.

3.8.1 [Optional] Assignment: Configure Bitwarden Two-Step Authentication

In this assignment, you enable Bitwarden 2-Factor Authentication for a website. This is normally referred to as 2FA, and called 2-Step Authentication, 2-Step Verification, Multi-Factor Authentication, and in the case of Bitwarden *Authenticator Key (TOTP)* for *Time-based One-Time Password)*. For this assignment we use Bitwarden to setup two-step authentication for a Google account.

---

[21] *https://bitwarden.com*

- Prerequisite: Installation of Bitwarden (performed in an earlier assignment) and purchase of one of its premium tiers.

1. Open a browser and go to *https://security.google.com.*

2. If prompted, authenticate.

3. In the main body area, scroll down to tap on *2-Step Verification.*

4. At the prompt, enter your Google credentials to authenticate again. Then tap *Next.*

5. Scroll down to the *Authenticator app* section. Then tap *SET UP.*

6. At the *Get codes from the Authenticator app,* select the type of device you use (Android or iOS), then tap *Next.*

7. The *Set up Authenticator* window is designed to be captured with a smartphone camera. As we are using a computer, tap *CAN'T SCAN IT?*

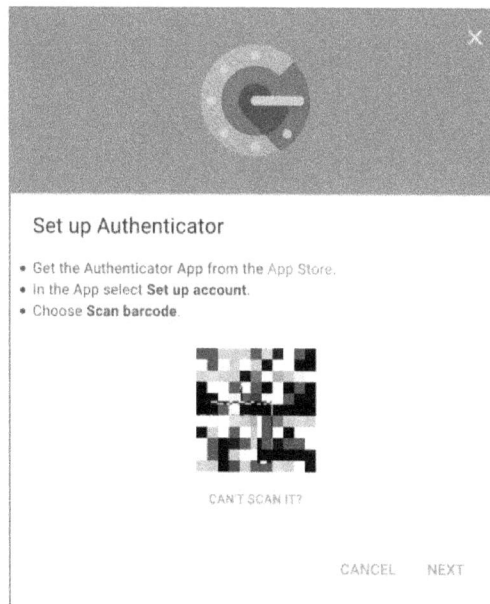

8. In the *Can't scan the barcode?* dialog, select then copy the 32-character code.

9. Open *Bitwarden,* select the account for two-factor authentication (Google), then tap the *Edit* (pencil) icon.

10. Paste the code copied in step 8 into the *Authenticator Key (TOTP)* field. Then tap the *Save* (disk) icon.

11. In the *ITEM INFORMATION* area of your Bitwarden Google record, you now see a *Verification Code (TOTP)* field. This is the one-time-only authenticator code that can be used when prompted by Google. If you have other devices with Bitwarden, they now also have this new field.

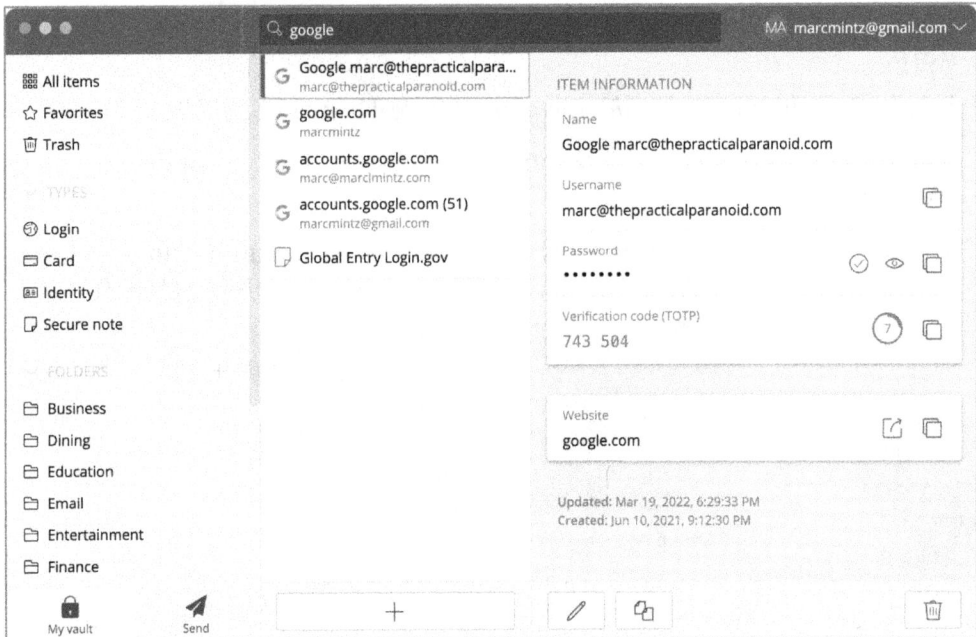

12. Return to the security page for the target site. It will ask for the 2FA code from your authenticator app. Paste the code into the site page:

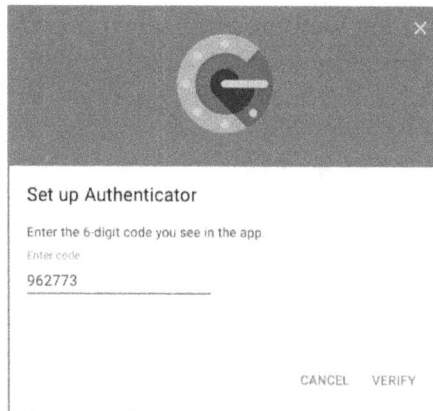

13. Tap *Verify*.

14. The site will display a confirmation page.

Enter a 2FA Code from Bitwarden

15. Open Safari to the website you just configured 2FA.

16. Authenticate with your username and password.

17. When prompted for Two-Factor Authentication, tap the Bitwarden icon to drop down the Bitwarden pane.

18. Tap the *clock* icon to the far right of the site name in the Bitwarden pane. This copies the 2FA.

19. Tap in the browser 2FA field, then *paste*. The 2FA is entered in the field.

20. Done!

3.8.2   Assignment: Configure 2FA in Passwords System Setting

In this assignment, you configure 2FA using *Passwords System Setting*.

- Prerequisite: macOS 12 and higher.

Configure 2FA in Passwords System Setting

1. Open *System Settings > Passwords*.

2. Authenticate.

3. From the *Passwords* main window area, select the site that you want to add 2FA protection. For this assignment, I've selected one of my Gmail accounts.

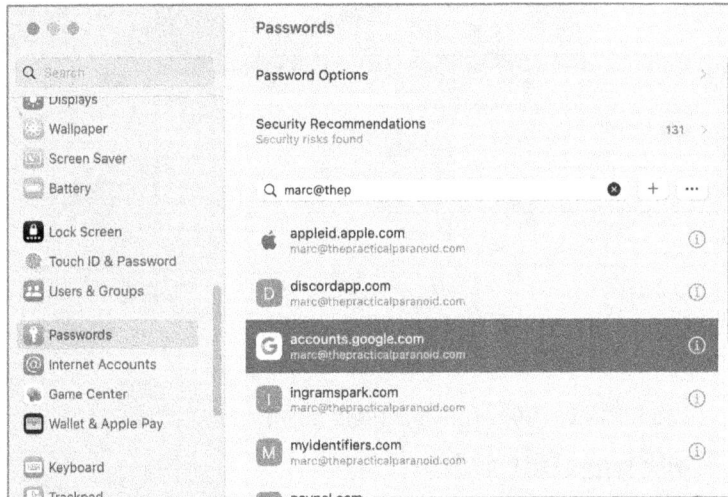

4. Double tap the target item.

5. Scroll down to *Verification Code*, then tap *Setup* button:

6. The *Set Up Key Verification Code* window opens:

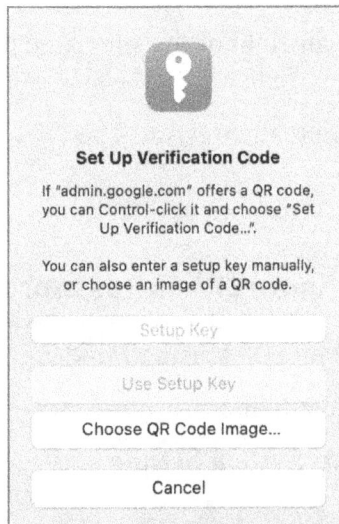

7. Open a browser to the website security page you want to configure 2FA. In this example, that is *https://myaccount.google.com*.

8. Navigate to the section to set up 2FA or Authenticator Keys. In this example, Google presents a scanning code. I tap *Can't Scan It:*

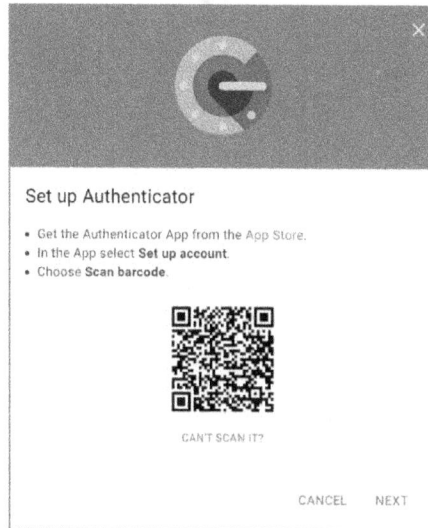

9. The *Can't scan the barcode?* Window opens, displaying the Verification Code.

10. Select and copy the Verification Code.

11. Tap the *Next* button. Keep this window open.

12. Return to the *Passwords System Preference* window, and then paste the code into the *Setup Key* field.

13. *Passwords* now displays the 2FA *Verification Code* for the site. This code will change ever 30 seconds.

14. Tap on the Verification Code to copy it.

15. Return to the security page for the target site. It will ask for the 2FA code from your authenticator app (*Passwords*). Paste the code into the site page:

Set up Authenticator

Enter the 6-digit code you see in the app.

Enter code

962773

CANCEL   VERIFY

16. Tap *Verify*.

17. The site will display a confirmation page:

Done!

You're all set. From now on, you'll use Authenticator to sign in to your Google Account.

DONE

18. Done!

Enter a 2FA Code from Passwords System Preference

19. Open Safari to the website you just configured 2FA.

20. Authenticate with your username and password.

21. At the prompt for Two-Factor Authentication, tap inside the code field. The option to autofill with the 2FA code appears. Select the autofill option.

- Note: If the autofill option does not appear, then:

    a. Open *System Settings > Passwords,* then select the target site from the sidebar.

    b. Tap on the *Verification Code* field > select *Copy Code.*

    c. Return to the site page.

    d. Paste in the verification code.

22. Done!

3.9   Passwords Lessons Learned

- ☐ Documents of NSA origin, acquired by Edward Snowden and leaked to The Guardian newspaper in 2013, provided evidence the NSA was legally and illegally spying on US citizens' cell phone, email, and web usage.

- ☐ Each website and service should have its own unique password.

- ☐ The first measure of a passwords strength is a minimum of 15 characters.

- ☐ The second measure of a passwords' strength is a mix of upper and lower case characters, numerals, and special characters.

- ☐ A unique password should be created for every site and service.

- ☐ Keychain is the macOS built-in password manager.

- ☐ Keychain stores its user database in *~/Library/Keychain.*

- ☐ Keychain Access is the utility to view the Keychain database.

- ☐ Keychain database can be shared and synchronized across all Apple devices sharing the same Apple ID account.

- ☐ Challenge Questions should not have Answers that are true.

- ☐ Bitwarden is a cross-platform password manager and Time-based One-Time Password (TOTP) Authenticator.

- ☐ Bitwarden requires a web browser extension to communicate with browsers.

- ☐ New as of macOS 12 is the ability to view and edit passwords from the *Passwords System Preference.*

- ☐ New with macOS 12 is the ability to import and export passwords (.csv format) from the *Passwords System Preference*

- ☐ A Password Policy is a set of rules to help users create and use passwords.

- ☐ Password Policies can be created for macOS using Apple Configurator, available free from the App Store.

- ☐ Two-Factor Authentication (2FA) resolves the vulnerability of cracked passwords.

- ☐ 2FA requires a secondary verification in the form of an email, SMS or text message, code generated by an application, or code generated by a hardware device.

- ☐ New as of iOS 15 and macOS 12 is the ability to share Keychain passwords with Windows users with iCloud for Windows, and iCloud Passwords browser extension for Chrome and Edge.

- ☐ New as of macOS 12 is built-in 2FA in the *Passwords System Preference.*

4 System and Application Updates and Privacy

Every new beginning comes from some other beginning's end.

–Seneca[1], Roman philosopher, statesman, and dramatist

What You Will Learn in This Chapter

- Configure Apple system and application update schedule
- Manually update OS and App Store apps
- Manage application updates with MacUpdater
- Configure system and application privacy
- Configure Gatekeeper

What You Will Need in This Chapter

- No additional resources required

4.1   System Updates

Most computer and mobile device users simply fail to update their systems. Many say the reason is that updates slow down the device, or they are concerned about introducing instability to their systems.

---

[1] *https://en.wikipedia.org/wiki/Seneca_the_Elder*

Updates rarely significantly change performance, although upgrades often do because they have a larger code base, requiring more RAM and CPU. While it is occasionally true that updates introduce instability, it is far more likely that updating creates greater stability.

More important is that many updates and upgrades patch vulnerabilities and security holes in a system. Fixing these security issues is so important that US-CERT (Homeland Security division responsible for cyber terrorism and IT security) strongly recommends that all users update all computers and mobile devices "as soon as possible."[2, 3, 4]

There are fundamentally three reasons for updates and upgrades:

- **Bug fixes**. All software and hardware have bugs. We never will be rid of them. Developers do want to squash as many as possible so that you are so happy with their product and continue to pay for upgrades.

- **Monetization**. Updates to operating systems and applications are usually free. Upgrades are for fee. But developers may include significant new features in an upgrade to encourage the market to purchase. Purchases are necessary so developers can afford to stay in business.

- **Security patches**. Although rarely talked about, one of the most important reasons for an update is to patch newly discovered security holes. Without the update, your computer may be highly vulnerable to attack today even if yesterday everything was secure.

It is for this last reason alone that I implore clients to be consistent with the update process. To protect your computer from security holes, it is critical to check for operating system and application updates daily. Fortunately, we can automate this process.

The typical macOS computer may have well over 500 applications, plug-ins, drivers, etc. The macOS, Apple applications, and apps downloaded from the Apple App Store can be updated through the App Store app and Software Update System Setting. Some applications have built-in automatic updating, but it is not

---

[2] *https://www.us-cert.gov/ncas/tips/ST04-006*

[3] *http://nvlpubs.nist.gov/nistpubs/SpecialPublications/NIST.SP.800-167.pdf*

[4] *https://www.cisecurity.org/critical-controls/documents/TheASD35andCISControls.pdf*

the norm and most system preferences, plug-ins, and other software do not automatically update. The easiest and most cost-effective way I've found to keep software up to date is to automate the process using MacUpdater[5].

macOS only allows the installation of applications that are downloaded from the Apple App Store, or from a registered and identified developer. Restricting applications from just the App Store is the safer option but is probably too restrictive for most environments. The built-in *Gatekeeper* enforces this restriction, and can be configured from *System Settings > Security & Privacy > General* tab.

New with macOS 13 is the ability of Apple to automatically push important security updates to the device, without the need for an administrator of the device to authorize the update.

4.1.1 Assignment: Configure System and Application Updates

In this assignment, you automate the process of updating the macOS operating system, as well as Apple software.

1. Open *Apple* menu > *System Settings* > *General* > *Software Update*.

2. Tap the *Automatic updates* ⓘ icon.

3. Enable all options, then tap *OK*:

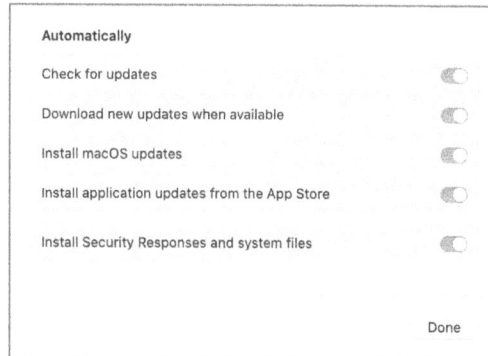

- Enable *Check for updates.*

---

[5] *https://www.corecode.io/macupdater/*

- Enable *Download new updates when available.* With this option active, the updates are downloaded in the background. Then an alert appears telling you the updates are ready to be installed. Installation starts immediately upon you tapping *OK* or *Install.*

- Enable *Install macOS updates.* You always want to have the most current macOS installed.

- Enable *Install app updates from the App Store.* It is vital to keep all your apps up to date.

- Enable *Install system data files and security updates.* This action completes the update steps.

4. Close System Settings.

4.1.2   Assignment: Manually Update OS and App Store Apps

You have the option to manually update OS and App Store apps.

In this assignment, you discover how to manually update OS and App Store apps, and to download App Store apps.

1. Select *Apple* menu > *App Store.* The Apple App Store opens.

2. From the sidebar, select *Updates* to view any available updates.

3. If updates are available, select their *Update* buttons. They download and automatically install.

4. To browse the App Store, select the *Discover, Create, Work, Play, Develop,* or *Categories* links in the sidebar.

5. Quit the App Store application.

Your macOS system and App Store applications now automatically alert you when updates are available.

4.1.3   Assignment: Install and Configure MacUpdater

MacUpdater is an application to helps keep all your applications, plug-ins, and drivers up to date.

114

In this assignment, you download, install, and configure MacUpdater free trial. The trial version operates 10 times at no cost.

1. Open a browser to the *MacUpdater* home page at *https://www.corecode.io/macupdater/*. Select the *Download* button at the bottom of the page.

2. Once the installer downloads, double tap to open it.

3. Drag the MacUpdater icon from the installer window to the Applications folder

4. Launch MacUpdater.

5. At the *Preferred Mode* screen, select *Use Background Mode.*

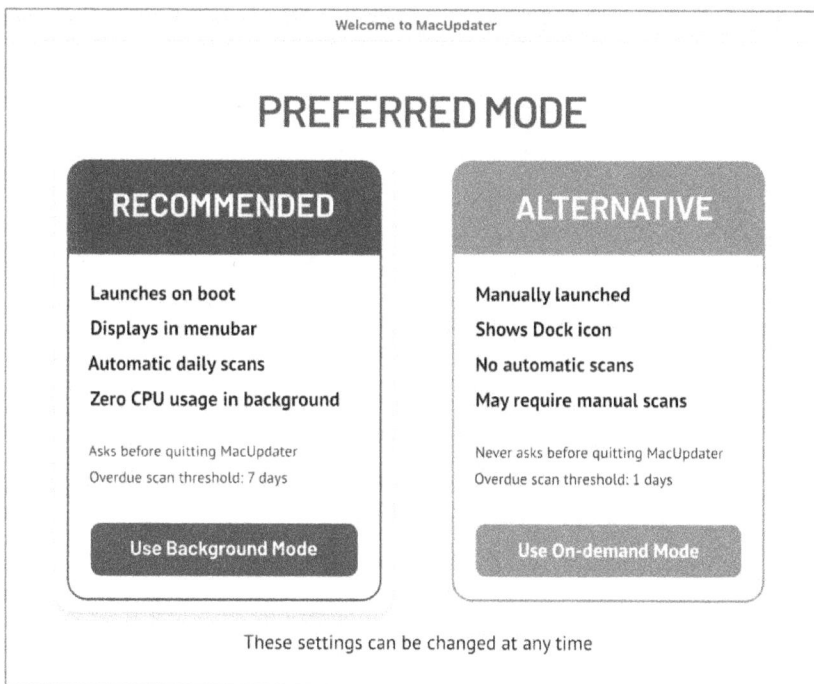

6. At *The Basics* screen, take the tour to learn about MacUpdater.

7. Once MacUpdater opens, it automatically starts to scan applications.

8. Once the scan completes, select the *Settings* icon in the tool bar, then configure to your taste. My settings are shown below:

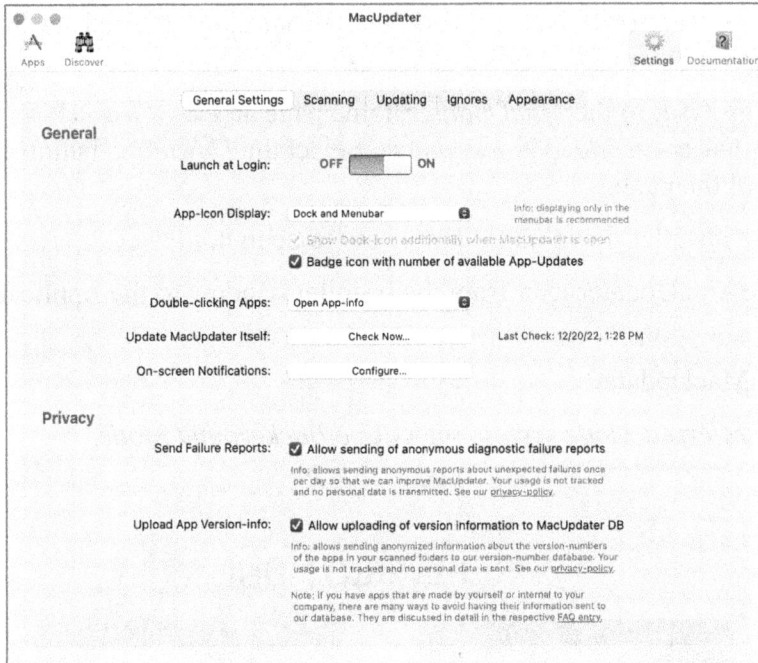

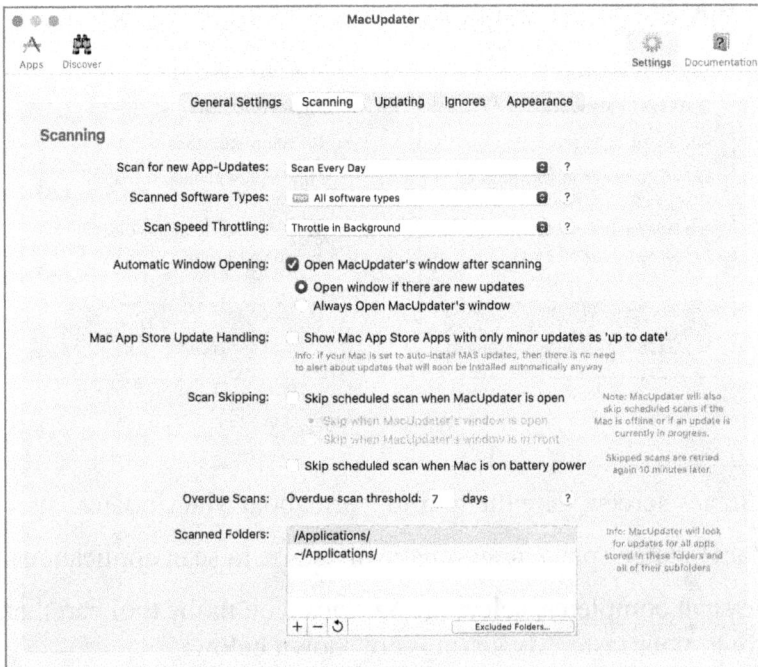

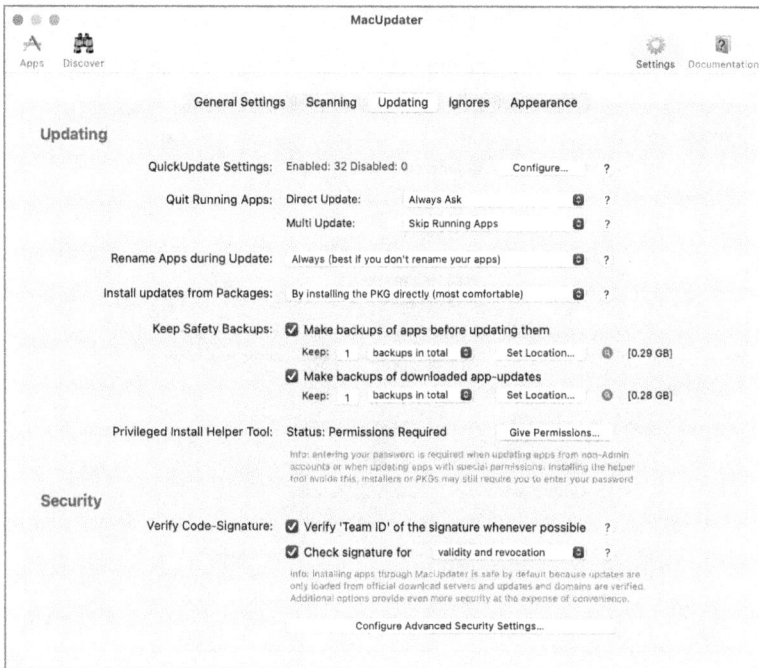

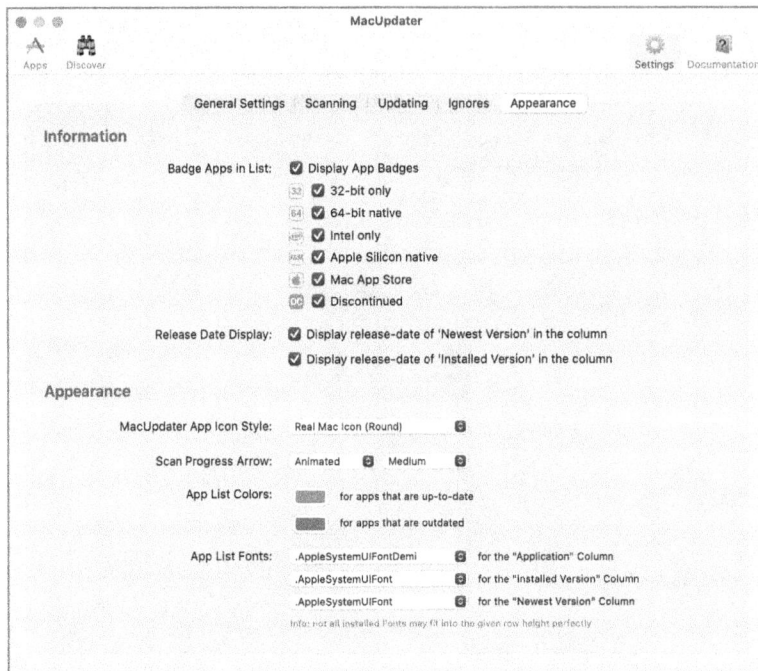

9. Tap the *App* icon in the tool bar.

10. Minimize MacUpdater as it can only perform auto-updates when running.

11. You have successfully installed and configured MacUpdater on your computer.

12. You can use MacUpdater for free up to 10 times. MacUpdater offers a link from within the application to securely purchase the application.

4.1.4 [Android] Assignment: Update all Apps

4.1.5 [Android] Assignment: Require Authentication for App Purchases

4.2 System and Application Privacy

System and Application developers often need feedback on how their users use their product. However, we users must first be respectful of our own privacy and give a thumbs-up or down to letting the system, application, and developer look over our shoulders.

4.2.1 Assignment: Configure System and Application Privacy

In this assignment you configure the Privacy settings of your device. There are no correct set of settings, and each person needs to determine the proper balance of privacy/flexibility/functionality that works for them.

1. Open *System Settings > Privacy & Security.*

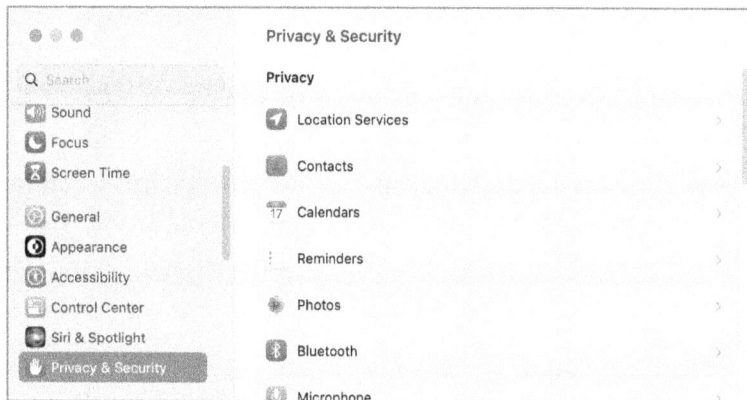

2. From the main area, select *Location Services.*

3. Enable *Location Services.*

4. Review all apps that want to know your GPS location. Many applications, such as Maps, need this information to work. But some apps do not. For example, I use an app called *Tone Gen* to produce pure tones in my audio engineering work. I cannot imagine why it needs to know my location, so I have it unchecked.

5. Tap the *Back* icon to return to *Privacy & Security.*

6. Repeat this process for each item in the main area.

    o Pay special attention to any items found when selecting *Full Disk Access.* Although many applications require full disk access to function properly, part of Apple's built-in security protocol is to limit an application's disk access. If you are unsure if an application should have full disk access, uncheck it. If it no longer functions correctly, enable it.

    o Pay special attention to *Analytics & Improvements > Improve Siri & Dictation.* If this is enabled, Apple has full access to all your Siri and Dictation recordings regardless of content.

7. When complete, quit System Settings.

4.2.2 Assignment: Configure Gatekeeper

In this assignment, you configure Gatekeeper to your taste.

1. Open *System Settings > Privacy & Security.*

2. Scroll down to the *Security* area > *Allow apps downloaded from:* area, then select either:

    o *App Store* if you wish to allow only applications downloaded from the App Store to run on the computer (most secure, but somewhat limiting), or

    o *App Store and identified developers* if you wish to allow any authorized application to run on the computer (potentially less secure, but less limiting.)

3. Quit *System Settings.*

4.2.3 [Android] Assignment: Clear Your Local Search History in Play Store

4.2.4 [Android] Assignment: Secure Play Store from Harmful Apps

4.3 System and Application Updates Lessons Learned

☐ System and application updates are an essential part of security and privacy.

☐ New with macOS 13 is the ability of Apple to automatically push important security updates to the device, without the need for an administrator to authorize the update.

☐ System and application updates exist to monetize the product, implement bug fixes, and eliminate potential and existing security vulnerabilities.

☐ The OS and Apple software update configuration is accessed from *Apple* menu > *System Settings* > *Software Update* > *Advanced.*

☐ App Store apps are updated from *Apple* menu > *App Store.*

☐ MacUpdater is a third-party utility that helps to automate updating applications from Apple, the App Store, and other sources.

☐ System and application privacy settings are accessed from *System Settings* > *Security & Privacy* > *Privacy* tab.

☐ Gatekeeper is a built-in security module, accessible from *System Settings* > *Security & Privacy* > *General* tab, that restricts running applications to only those from the Apple App Store or registered and identified developers.

4.4   Additional Reading

Souppaya, Murugiah, and Karen Scarfone. "Guide to Enterprise Patch Management Technologies." <u>NIST Special Publication 800-40, Revision 3</u>. July 2013. <http://nvlpubs.nist.gov/nistpubs/SpecialPublications/NIST.SP.800-40r3.pdf>

Liu, Simon, Rick Kuhn, and Hart Rossman. "Surviving Insecure IT: Effective Patch Management." <u>Insecure IT</u>. 2009. <http://csrc.nist.gov/staff/Kuhn/liu-kuhn-rossman-v11-n2.pdf>

5 User Accounts

Those who desire to give up freedom in order to gain security will not have, nor do they deserve, either one.

–Benjamin Franklin[1]

What You Will Learn in This Chapter

- Understand the 6 types of user accounts
- Never log in as an administrator
- Enable root user
- Log in as root user
- Change root user password
- Disable the root user
- Create an administrative user account
- Change from administrator to standard user
- Configure screen time control and application whitelisting
- Create a policy banner

What You Will Need in This Chapter

- No additional resources required.

---

[1] *https://en.wikipedia.org/wiki/Benjamin_Franklin*

5.1   User Accounts

User accounts allow you to compartmentalize your data so that it is not exposed to those who may be using your device. You may end up doing all the other security steps to harden your device to the core, but otherwise your data could be exposed when your friend uses your device for 5 minutes to "check my email." If you have the same users commonly using your device, then it is a very good idea to assign them their own user accounts. By assigning a user a separate user account, that user cannot access your specific data such as pictures, app-specific data, etc. Each new user account has its own authentication method (or none if the user chooses), separate from your own method.

The user accounts feature is of great use for parents or co-owners of a device such as spouses. Separating your workspaces allows for customization and convenience in how the device operates for each user. Also, having your data only within your workspace and locked behind your authentication is critical to protect against accidental data compromise.

Types of User Accounts

macOS allows five different types of user accounts, plus the ability to monitor and control a user account called *Screen Time Controlled*). Each account type has its own pros and cons, powers, and limitations. All but one–Root–you may create and modify from the *Users & Groups System Setting*.

- **Root**. The root account is created automatically upon system installation and cannot be created or deleted any other way. There can be only one root account on any computer. Root is the ultimate lord over the system, with unquestioned power and control. If root does something dangerous–for example, issues a command to erase the entire drive–the system does not even issue a *Danger, Will Robinson* alert, it simply dutifully erases the drive. Root is present out of the box but is disabled by not having a password assigned. It is rare to ever need to enable the root user, as any administrator account can assume the powers of root.

- **Administrator**. There must always be at least one administrator. There may be an unlimited number of administrators, or administrative user accounts, each having identical power over the computer. What makes an administrator unique above the standard, sharing only, and guest user accounts are its abilities to:

o Create new user accounts and delete user accounts

o Modify the contents of some restricted folders (Library, Applications)

o Authorize the installation or removal of applications and system updates

o Take on the powers of root from the command line by issuing *sudo* and *su* commands.

• **Standard**. There can be as few as zero, and up to an unlimited number of, standard accounts. Standard is the recommended account level for most users working locally on the computer. Standard accounts can open and work without limitations using any application installed on your Mac. The advantage of working as a standard account is that it is not possible to damage the operating system or applications.

• **Screen Time Controlled**. Screen time is not a type of user account, it is a monitor and control over accounts. This account is typically a standard account. Screen time can restrict the powers of the account by limiting access to specific applications, access to specific websites or any adult site, who can communicate with the user via Apple Mail and Messages, the hours for which the user may stay logged in, etc. Although this control was originally intended to protect children from the darker areas of the Internet and protect the computer from the children, it is a powerful tool for use with employees (guess how many billions of dollars a year in wasted productivity are spent on Facebook? At least $3.5 *Trillion US[2]*).

• **Sharing Only**. You can have from zero to an unlimited number of sharing only accounts. This type of account cannot log in locally to the computer. The only access is via the network and file sharing. A sharing only account is useful if you need to work and share files with someone on the same network. This allows them to access your computer and specific files over the network.

• **Guest**. There is only one guest account. With guest enabled, anyone may access your computer as a guest, either locally or via file sharing over the network. The guest only has access to folders and files that are shared as

---

[2] *https://www.nbcnews.com/tech/social-media/time-wasted-facebook-could-be-costing-us-trillions-lost-productivity-n511421*

either read or read and write for *everyone*. If a guest logs in locally, any documents the guest creates and saves in the guest home folder are deleted upon log off. Unless you are certain of your file-sharing configuration, it is insecure to allow guest access.

5.2 Never Log In as an Administrator

It is the human condition: We want power, authority, and more power! This carries over into how we log in the computer. Everyone wants to be the administrator of their computer! Apple enables this. When the owner of a new Mac boots up for the first time, that person is prompted to create a user account, which is by default an administrator account.

But this is bad practice. If you have the bad luck of launching malware on your computer (most often unknowingly) while you are logged in as an administrator, the malware takes on your user account power. This means the malware has full control and power over the computer, including all user accounts. Yikes.

On the other hand, if you launch malware while logged in as a non-administrative user, the malware typically takes on your non-admin power. Under this scenario, the malware has full control over your home folder and nothing else.

I can hear the wailing from here: "But I need to be an administrator. How else am I able to install software and updates, and perform maintenance?"

Fear not. You do not need to be logged in as an administrator to perform administrator tasks (adding/deleting user accounts, installing/updating the system and applications, and running system diagnostic and repair utilities). You can be logged in with any type of user account. You only need to authenticate with an administrator name and password when prompted.

To do this, you need to have an administrative user account on the computer but log in with a non-admin (standard) user account. When you are prompted for an admin name and password while performing admin duties as a non-admin, just enter the admin name and password.

5.2.1 Assignment: Enable the Root User

Before jumping in and enabling root by assigning a password, give thought to why you want to enable root. Any administrative account can assume the powers of root whenever needed. I have seen far too many users send their data to the cornfield by logging in as root then making a simple keystroke error.

There is one condition that requires enabling root–allowing unattended remote access that requires a reboot.

However, if you wish to experiment with root powers, here we go:

1. Select *Apple* menu > *System Settings* > *Users & Groups*.
2. Authenticate.
3. Select the *Login Options* button.
4. Select the *Network Account Server: Join* or *Edit* button.
5. Select the *Open Directory Utility...* button.
6. Tap the lock icon, then authenticate with administrator credentials.
7. Select the *Edit* menu > *Enable Root User*.
8. In the *Please enter a new password for the root user* window, enter a strong password, verify, then tap the *OK* button.
9. Quit Directory Utility.
10. Quit System Settings.

Root was on the computer from the moment the system was installed. Giving root a password enabled it.

5.2.2 Assignment: Log In as Root

When logged in as the root user, you have full read/write access to most otherwise restricted areas, such as every users' home folder.

In this assignment, you log in as the root user.

*   Prerequisite: Completion of the previous assignment.

1. Log out of the current user account.

2. At the Login Window, log in as *root.* If you do not see the *root* user, select *Other...* From here you may enter the username `root` and the password you assigned for root.

3. Once at the Desktop, navigate to the */Users/<username>* folders. Notice that you can access any user folder with read and write permissions.

4. To log out, select the *Apple* menu > *Log Out.*

5. At the *Login Window,* log in with your standard account.

5.2.3   Assignment: Change Root Password

In this assignment, you change the root user password.

1. Select *Apple* menu > *System Settings* > *Users & Groups.*

2. Authenticate.

3. Select the *Network Account Server Join* or *Edit* button.

4. Authenticate.

5. Select the *Edit* menu > *Change Root Password.*

6. Enter a strong password.

7. Quit Directory Utility.

8. Quit System Settings.

5.2.4   Assignment: Disable Root

In this assignment, you disable the root user account.

The only factor that enabled root is having a password. If the password for root is changed to <null> (nothing), root is disabled.

1. Select *Apple* menu > *System Settings* > *Users & Groups.*

2. Authenticate.

3. Select the *Network Account Server... Join* or *Edit* button.

4. Select the *Open Directory Utility* button.

5. Authenticate.

6. Select the *Edit* menu > *Disable Root User.*

   - Note: The only thing this function performs is delete the password for root.

7. Quit Directory Utility.

8. Quit System Settings.

5.2.5 Assignment: Create an Administrative User Account

One of the most important rules in IT security is to log in with a non-administrative account, not an administrative account. However, the very first account created when you initially boot up your computer *is* an administrator!

In this assignment, you create an administrative user account on the computer. In the next assignment, you change your own account to a standard user account so that you are no longer be in violation of this administrator rule.

1. Log into the computer with your normal (administrator) account.

2. Open *Apple* menu > *System Settings* > *Users & Groups.* Tap the *Lock* icon in the bottom left corner, then authenticate with an administrator name and password.

3. Tap the + *(add user)* button at the bottom of the side bar. The *Create a New Account* window opens.

New Account:	Administrator ⊕
Full Name:	Administrator
Account Name:	administrator
Password:	••••••••••••••••
Verify:	••••••••••••••••
Password Hint: (Recommended)	Hint (Recommended)
?	Cancel Create User

   - From the *New Account* pop-up menu, select *Administrator.*

- In the *Full Name* field, enter `Administrator`.

- In the *Account Name* field, enter `administrator`.

- In the *Password* field, enter a strong password. For this course, write down this password. In the real-world environment, you must secure the written record of all passwords.

- In the *Verify* field, reenter the strong password.

- I do not recommend entering anything in the *Password Hint* field, as this can be helpful to hackers, not just you.

4. When done, tap the *Create User* button. You are returned to the *Users & Groups* preference.

5. *Quit* System Settings.

You have successfully created a new administrator account.

5.2.6   Assignment: Change from Administrator to Standard User

In the previous assignment, you created an administrative user account whose name and password can be used when needed.

In this assignment, you change your own account to a standard user account which remains your regular log in account.

1. Log out of your account.

2. Log in as the new administrator account.

3. Select the *Apple* menu > *System Settings* > *Users & Groups*.

4. Unlock the *Lock* icon, then authenticate with administrator.

5. Select your account in the side bar.

6. Disable the *Allow user to administer this computer* check box.

7. At the prompt informing the change takes place after a restart, select the *OK* button.

8. Select *Apple* menu > *Restart*.

9. Log in with your everyday account (now a Standard account).

Whenever you need to perform administrative tasks, use the name and password of the new administrator account you have just created. No need to log in as an administrator!

5.2.7   [Android] Assignment: Configure or Remove a User Account

5.3   Screen Time Control and Application Whitelisting

In 2014, Target, Home Depot, and other major retailers were hacked for their customer databases. Although there were multiple breakdowns in the security protocols of these organizations, one step would have prevented all of them: *application whitelisting.* This same strategy should be used by both home and business systems to help secure computer systems.

Application whitelisting is a process that allows only authorized applications to run on a computer, blocking any executable not on the list. This is a vital ingredient to system security because even the best anti-malware catches only 99.9% of the *known* bugs. And if your computer is penetrated by *unknown* malware, anti-malware is of no use. However, if your computer has application whitelisting in place, the unknown malware is blocked from executing.

macOS has built-in application whitelisting, limiting the execution of applications from either *only* the App Store, or from only the App Store and identified developers. This by itself almost eliminates the possibility of rogue code from running on macOS. This is part of the *Gatekeeper* endpoint protection[3].

In macOS, *Screen Time* (named *Parental Control* prior to macOS 11) can be used to perform application whitelisting. Screen Time allows an administrator to restrict access to specific applications and services provided to a user account. In the previous incarnation as *Parental Control*, this feature was intended as a way for parents to better manage their children's accounts. It also has its place in the business setting by restricting specific applications (disallowing Spotify, etc.),

---

[3] *https://support.apple.com/guide/security/app-security-overview-sec35dd877d0/1/web/1*

restricting access to specific websites (pornography, Facebook, etc.), allowing access to the account only during work hours, and monitoring computer use.

Once Screen Time has been used to implement application whitelisting it is necessary for the administrator to be available for a brief time while the unintended consequences shake out. It is common for some permitted applications to require the use of a restricted application or process. An administrator must be available to provide authorization.

5.3.1 [Android] Assignment: Enable Family Link

5.3.2 [Optional] Assignment: Configure Screen Time

For this assignment, you configure your own account to add the security of application whitelisting. These same steps should be taken for all non-administrative accounts on your computer and all computers in your household or business. Best practices hold that *all* non-administrative accounts should have application whitelisting enabled–and you never log in with an administrative account.

Setup Family Sharing

This part is optional. If you would like to manage up to six family members, complete this part. Otherwise, skip to the next section *Enable Screen Time*. More information on Family Sharing can be found at *https://support.apple.com/en-us/HT201088*. More information on Screen Time can be found at *https://support.apple.com/en-us/HT210387*.

1. Log in with a non-administrator user account.

2. Open *Apple* menu > *System Settings* > *Apple ID* > *Family Sharing* icon > *Set Up Family* button.

3. In the *Invite people to your family* window, you may invite family members, create a Child Account, or select *Not Now*. For this assignment, tap *Not Now button.*

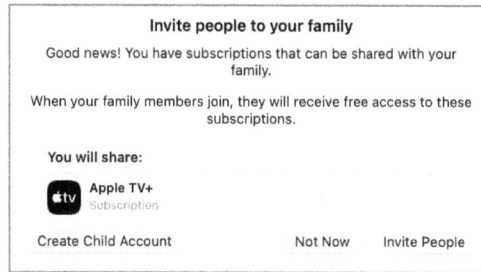

- To invite family members to join the *Family Sharing,* or to enable Family Sharing on an iOS device, view the document at *https://support.apple.com/en-us/HT201088.*

Enable Screen Time

4. Open *System Settings > Screen Time.*

5. If you have not enabled Family Sharing, skip to the next step. If you have enabled Family Sharing, select your target user from the user menu directly under your picture.

6. Tap the *Options* button at the bottom of the sidebar. Configure as below:

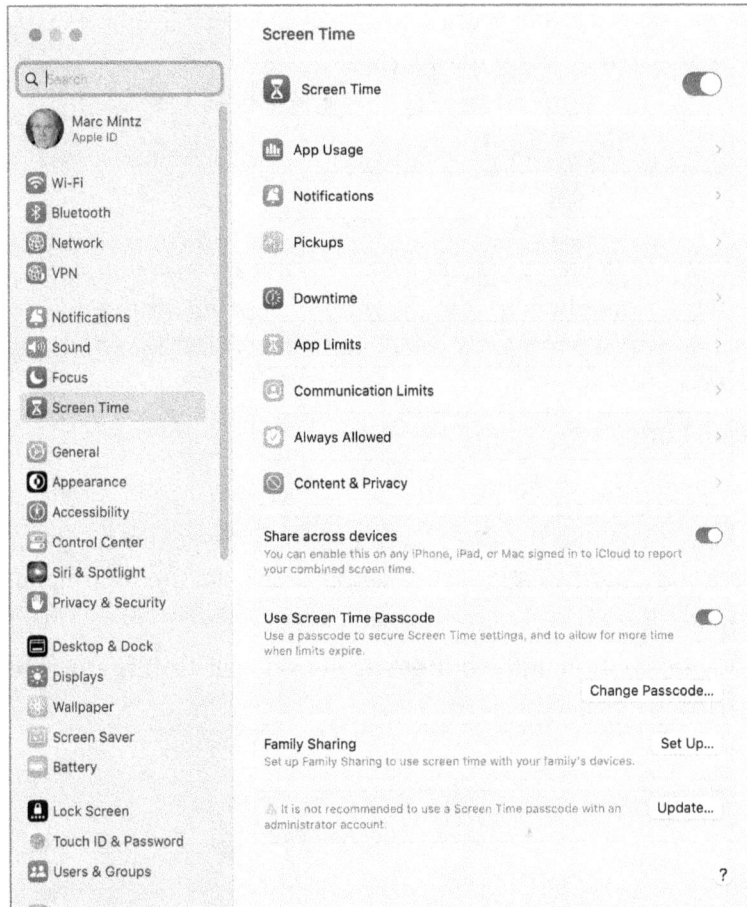

a. Tap the *Turn On* button to enable Screen Time.

b. Enable the *Share across devices* checkbox to be able to see usage information for all Macs signed into iCloud with this user.

c. To be able to see usage information from iPhone, iPad, and iPod devices, on each device go to Settings > Screen Time, and enable the same setting.

Set Up Screen Time Passcode

Without a passcode, Screen Time can be modified or disabled by the user. This may work well for your own account but may not be the best security option for a child or employee.

With a passcode active, any changes to the users Screen Time or time extensions to restricted applications require entry of this passcode.

If using Family Sharing to manage an account, follow these steps. If not using Family Sharing, skip to step 13.

7. Select *Apple* menu > *System Settings* > *Screen Time.*

8. Select the target user account from the menu in the upper-left corner.

9. Tap *Options...* at the bottom of the sidebar.

10. Select *Use Screen Time Passcode*, enter a passcode, then verify the passcode.

11. If prompted to enter your Apple ID for Screen Time passcode recovery, enter your Apple ID and passcode. Then skip to step 16.

If you are not using Family Sharing:

12. Using the computer that has the target user account, log into the target user account.

13. Select *Apple* menu > *System Settings* > *Screen Time* icon.

14. Select *Use Screen Time Passcode,* enter a passcode. Then verify the passcode. Write down the passcode. In a real-world environment, securely store the passcode.

15. Configure *Downtime, App Limits, Communication, Always Allowed,* and *Content & Privacy* windows to your own needs.

16. Quit System Settings.

5.3.3 [Android] Assignment: Turn On and Configure Play Store Parental Controls

5.3.4 [Android] Assignment: Enable Pin Windows

5.3.5 [Windows] Assignment: Configure a Child Account

5.3.6 [Windows] Assignment: Allow Access to Websites Through Email

5.4 Policy Banner

Within some organizations, the legal or IT department specifies there must be a *Policy Banner* present at startup. This alerts any would-be hackers or criminals that proceeding into the computer is a criminal offense. Having a policy banner in place may prevent the "I didn't know I was doing anything wrong" defense in court. It is also a requirement of the NIST 800-171 and ISO 27001.

5.4.1 Assignment: Create a Policy Banner

In this assignment, you create a policy banner that displays upon startup.

Create the Policy Banner file

1. Open a word processor or text editor that can create a plain text (.txt) or rich text (.rtf) file format.

- Note: For this example, I create an .rtf file.

2. Create a new document with the specifics required for your policy banner. A sample policy banner is listed below, provided by the United States Cybersecurity & Infrastructure Security Agency[4] (CISA):

```
*** WARNING ***


By clicking [ACCEPT] below you acknowledge and
consent to the following:
```

---

[4]*https://www.cisa.gov/sites/default/files/publications/Nine%20Elements%20of%20Consent_Private_Sector_091620.pdf*

- All communications and data transiting, traveling to or from, or stored on this system will be monitored.

- You consent to the unrestricted monitoring, interception, recording, and searching of all communications and data transiting, traveling to or from, or stored on this system at any time and for any purpose by [the COMPANY] and by any person or entity, including government entities, authorized by [the COMPANY].

- You also consent to the unrestricted disclosure of all communications and data transiting, traveling to or from, or stored on this system at any time and for any purpose to any person or entity, including government entities, authorized by [the COMPANY].

- You are acknowledging that you have no reasonable expectation of privacy regarding your use of this system.

- These acknowledgments and consents cover all use of the system, including work-related use and personal use without exception.

3. Save the file as *PolicyBanner* in either .txt or .rtf, to the Desktop.

4. Drag and drop the Policy Banner file into the */Library/Security* folder.

5. At the prompt, enter your administrator credentials to authorize the copy.

6. Open Terminal to adjust permissions so that Everyone (Other) has read and execute privileges.

7. If you are logged in with a non-administrator account (as you always should be), log into Terminal as an administrator:
   `su <administrator name>` then tap Return.

At the prompt to enter a password, enter the administrator password, then tap Return.

8. Enter on one line:

* For .txt:
  `sudo chmod o+rx /Library/Security/PolicyBanner.txt`

* For .rtf:
  `sudo chmod -R o+rx /Library/Security/PolicyBanner.rtf`

9. At the prompt, enter the administrator password.

10. Quit Terminal.

Test the Policy Banner

11. From the Apple menu, select *Restart*.

12. When the *Policy Banner* appears, tap the *Accept* button.

* NOTE: This policy banner is in .rtf format. This allows for text formatting.

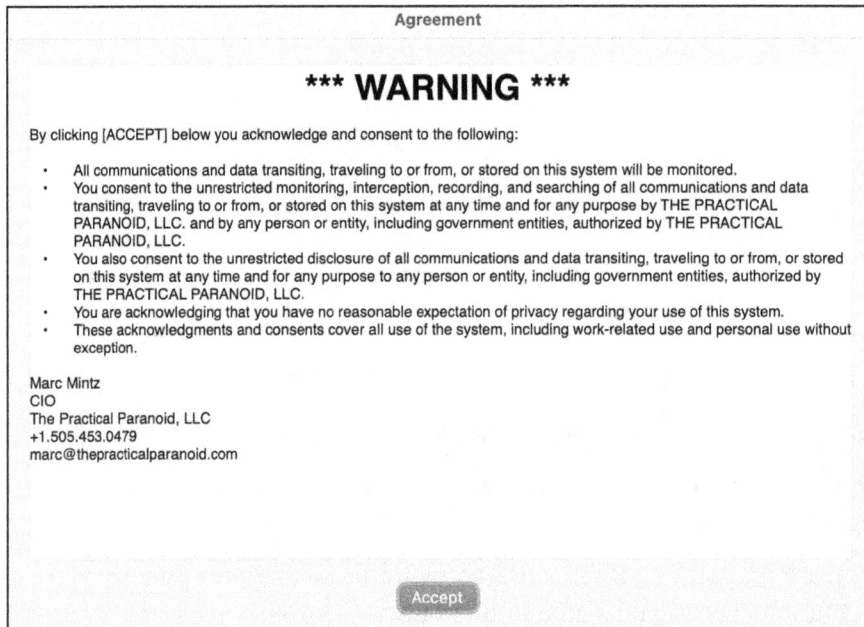

Agreement

*** WARNING ***

By clicking [ACCEPT] below you acknowledge and consent to the following:

* All communications and data transiting, traveling to or from, or stored on this system will be monitored.
* You consent to the unrestricted monitoring, interception, recording, and searching of all communications and data transiting, traveling to or from, or stored on this system at any time and for any purpose by THE PRACTICAL PARANOID, LLC. and by any person or entity, including government entities, authorized by THE PRACTICAL PARANOID, LLC.
* You also consent to the unrestricted disclosure of all communications and data transiting, traveling to or from, or stored on this system at any time and for any purpose to any person or entity, including government entities, authorized by THE PRACTICAL PARANOID, LLC.
* You are acknowledging that you have no reasonable expectation of privacy regarding your use of this system.
* These acknowledgments and consents cover all use of the system, including work-related use and personal use without exception.

Marc Mintz
CIO
The Practical Paranoid, LLC
+1.505.453.0479
marc@thepracticalparanoid.com

Accept

13. Continue log in as normal.

5.5   User Accounts Lessons Learned

☐ User accounts allow compartmentalization of your data, so the data is not exposed to others using your device.

☐ Each user of a device should have their own user account.

☐ The root account has absolute authority over the device.

☐ There can be only one root account.

☐ By default, root is disabled until a password is assigned to it.

☐ An administrator account is the default account for the first user of the device.

☐ There can be an unlimited number of administrator accounts.

☐ An administrator can create new user accounts, delete user accounts, modify the contents of some restricted folders, authorize the installation or removal of applications and system updates, and assume the power of root from the command line.

☐ Almost all administrator functions can be performed while logged in as a standard user by entering administrator credentials at the prompt. It is important to not log in with an administrator account unless there is no other reasonable option. When logged in as an administrator, any malware that compromises the system likely assumes the power of the logged in user account.

☐ Most users should be assigned standard accounts. Users of standard accounts cannot damage the operating system or add/delete applications.

☐ Screen time controlled is not a type of user account, but allows monitoring and some controls over the user account. Screen time controlled can limit access to specified applications, access to specified websites, who can communicate with the user via Apple Mail and Messages, and the hours for which the user may stay logged in, and more.

☐ Screen Time Controlled is enabled from *System Settings > Screen Time*.

☐ A sharing only account is designed for users who do not log into the device locally but need to access your device data remotely from their own devices. There can be a virtually unlimited number of sharing only accounts.

☐ There can be only one guest account. With guest enabled, anyone may logon to your computer without need for username or password. When the guest logs off, all data created is erased.

☐ It is exceptionally rare to ever need to enable the root user.

☐ To enable root, go to *Apple* menu > *System Settings* > *Users & Groups* > *Login Options* > *Network Account Server* > *Join* or *Edit* button > *Open Directory Utility* > *Edit* menu > *Enable Root User*.

☐ To change a user from administrator to standard account, once the user is logged off, log in with a different account > *System Settings* > *Users & Groups*> select the target user account > disable *Allow user to administer this computer* checkbox.

☐ Enable application whitelisting to almost entirely eliminate problems caused by malware.

☐ Create a policy banner that displays to users before they access the computer. This is a NIST and ISO requirement. Even if you have no need to comply with these requirements, having a policy banner in place may aid in prosecuting hackers for unauthorized access.

6 Device Hardware

I am disturbed by how states abuse laws on Internet access. I am concerned that surveillance programs are becoming too aggressive. I understand that national security and criminal activity may justify some exceptional and narrowly tailored use of surveillance. But that is more reason to safeguard human rights and fundamental freedoms.

–Ban Ki-moon[1], Secretary General of the United Nations

What You Will Learn in This Chapter

- Disable USB, FireWire, and Thunderbolt storage device access
- Boot into Target Disk Mode
- Boot into Recovery Mode
- Boot into Single-User Mode
- Enable FileVault 2
- Remotely access and reboot a FileVault drive
- Enable firmware
- Startup Security for Apple Silicon Macs

What You Will Need in This Chapter

- A second macOS computer.
- A Firewire, USB, USB-C, or Thunderbolt cable to connect two macOS computers.
- [Optional] External storage device of any size.

---

[1] *https://en.wikipedia.org/wiki/Ban_Ki-moon*

6.1   Block Access to Storage Devices

In some environments, it is appropriate to block access to external storage devices. This may be required so that users cannot copy sensitive data.

Prior to macOS 11, it was possible to block access to storage devices by renaming the drivers specific to the port used by the device. All these drivers are in */System/Library/Extensions*. However, with the introduction of macOS 11, Apple no longer allows tinkering under the hood.

Short of using epoxy to close off the ports (a common practice in some organizations), the only solution with macOS 11 and higher is the installation of *End Point Protection* software.

End Point Protection software is relatively expensive. If you need to go down this path, you may request demonstration versions of:

- DeviceLockDLP[2] by Acronis
- End Point Protector[3] by CoSoSys

6.1.1   [Windows] Assignment: Block Access to Storage Devices Via Registry

6.1.2   [Windows] Assignment: Restore Access to Storage Devices Via Registry

6.1.3   [Windows] Assignment: Block Access to Storage Devices Via Device Manager

---

[2] *https://www.devicelock.com*
[3] *https://www.endpointprotector.com*

6.1.4 [Windows] Assignment: Restore Access to Storage Devices Via Device Manager

6.2 Storage Device Encryption

FileVault

Strong passwords keep the network and Internet-based password attacks at bay, but should someone have physical access to your computer, they may be able to perform a brute force attack on your login password.

Prior to Mac OS X 10.7, the system included home directory encryption using *FileVault*, now referred to as *Legacy FileVault*. This was enabled on a user-by-user basis.

Starting with Mac OS X 10.7, we have FileVault 2[4], normally referred to as *FileVault*, which provides military-grade AES-256 full disk encryption. With FileVault enabled, your drive has a secure wall around it that can only be penetrated by entering an account password.

Enabling FileVault has an additional advantage: the boot time keyboard commands require authentication at the Login Window before proceeding. This is a vital security protection for three keyboard commands:

- **Target Disk Mode** (TDM) allows booting with the macOS functionally disabled and only USB, Firewire, and Thunderbolt active for storage devices. TDM thus effectively turns the computer into an external drive that can connect via USB, Firewire, or Thunderbolt.

- **Recovery Mode** allows booting into the otherwise invisible Recovery partition on the boot drive, so you may perform directory repair with Disk Utility, reinstall macOS, enable Firmware password, etc.

- **Single-User Mode** (Intel Mac only) allows booting into a command line state, prior to loading of Open Directory (the database holding all user accounts) and the entire OS. In this state, only the Root user account is active, hence the term *Single-User Mode*. In this mode, you can only type direct commands.

---

[4] *http://en.wikipedia.org/wiki/FileVault*

6.2.1 Assignment: Enable and Configure Storage Encryption

In this assignment, you enable full disk encryption on your boot drive using FileVault 2.

- Note: It is almost certain your macOS 13 boot drive was encrypted with FileVault 2 during the installation process. This can be verified if at step 3 the button reads *Turn Off FileVault...* If this is the case, skip this assignment.

1. Open *Apple* menu > *System Settings* > *Security & Privacy*, then scroll down to the *FileVault* area.

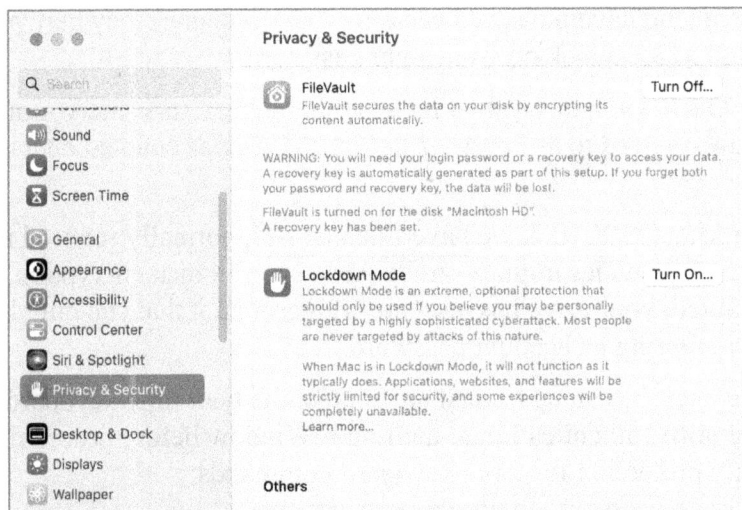

2. Select the *Turn On FileVault...* button.

3. A dialog box appears to select using either your iCloud account or a recovery key to unlock the disk in the event your login password is forgotten.

   - Selecting *Allow my iCloud account...* allows you to use your iCloud account password to be used, then select the *Continue* button.

   - Selecting *Create a recovery key...* presents a randomly generated password. Store this key in a secure location, I recommend in your Address Book / Contacts application. Then select the *Continue* button.

4. If there are multiple user accounts on this machine, you are asked to enable which user accounts are to be allowed to unlock the encrypted boot drive to boot up. For each of these accounts, tap the *Enable User...* button. Enter the

account password. Users that have not been enabled can still access their accounts via *Fast User Switching* after the drive is unlocked by one of the authorized accounts. Then tap *Continue*.

5. Select the Restart button to restart the Mac and begin the encryption process.

6. When your Mac returns to the Desktop, the *Security & Privacy* preference window reopens, providing a progress indicator for the encryption process. You may close this window if desired.

6.2.2   Assignment: Enable Encryption on Non-Boot Storage

Although the previous assignment works easily and quickly for encrypting your boot volume (and its associated data volume), the same process cannot be used for a standalone data volume (a non-boot volume).

The good news is that encrypting a standalone data volume is not much more than a 1-tap process. If there is data on the volume, there is no data loss.

In this assignment, you encrypt a non-boot volume. This may be a thumb drive, external hard drive, or SSD.

● Prerequisite: Possession of an external storage device.

1. Attach the drive to your computer.

2. *Right-tap* on the drive > *Encrypt* submenu.

3. At the prompt, enter a strong password, verify the strong password, then tap *Continue.*

4. Depending on the size and speed of the drive, encryption may take from minutes to days. If the computer goes to sleep or is shut down, encryption automatically continues once awake.

6.2.3   [Intel Mac] Assignment: Boot Into Target Disk Mode

In this assignment, you boot into Target Disk Mode (TDM) on an Intel Mac.

● Prerequisite: Use of an Intel Mac for TD, and an additional Intel or Apple silicon Mac. For Apple silicon Macs, please see the next assignment.

1. Connect two Mac computers, at least one of which is an Intel Mac. Use a FireWire or Thunderbolt cable.

2. Shut down the Intel Mac computer to be placed into TDM.

3. Power on the computer to be placed into TDM. Then immediately (before the appearance of the Apple logo) hold down the *T* key and keep it held down.

4. The login window appears. Release the T key.

5. Select your account, enter your password, then tap the Return/Enter key.

6. The Firewire or Thunderbolt icon appears, moving around your screen. You now are in TDM.

7. To verify you are in TDM, look at the second computer (power on if not already powered). Your TDM computer now mounts on the other computer.

8. To exit TDM, press and hold the TDM Mac computer power button to power off. Then power on as normal.

6.2.4   [Apple silicon Mac] Assignment: Boot Into Target Disk Mode

In this assignment, you boot into Target Disk Mode (TDM) on an Apple silicon Mac.

- Prerequisite: At least one Apple silicon Mac to be placed into TDM, and another Intel or Apple silicon Mac.

- Prerequisite: A USB, USB-C, or Thunderbolt cable to connect the two devices.

1. Connect two computers with a USB, USB-C, or Thunderbolt cable.

2. Shut down the TDM Apple silicon Mac.

3. Press and hold the TDM Apple silicon Mac power button until *Loading startup options* window appears.

4. Tap the *Options* button. Then tap the *Continue* button.

5. If prompted, enter an administrator account password.

6. The TDM Apple silicon Mac opens in Recovery mode.

7. Select *Utilities* menu > *Share Disk*.

8. Select the disk or volume to be shared. Then tap *Start Sharing.*

9. On the other Mac, open a Finder window, then tap *Network* (below Locations) in the sidebar.

10. In the *Network* window, double tap the TDM Mac that has the shared disk or volume. Tap *Connect As,* select *Guest* in the *Connect As* window. Then tap *Connect.*

11. When done sharing files, eject the disk of the TDM Apple silicon Mac.

12. Press and hold the power button of the TDM Apple silicon Mac to power it off.

6.2.5   [Intel Mac] Assignment: Boot Into Recovery Mode

In this assignment, you boot into Recovery Mode.

*   Prerequisite: Use of an Intel Mac.

1. Power off your Intel Mac computer.

2. Power on your Intel Mac computer, then immediately (before the appearance of the Apple logo) hold down the *cmd + R* keys and keep held down.

3. The Login Window appears.

4. Release the cmd + R keys.

5. Select your account, enter your password, then tap the Return/Enter key.

6. The Recovery home screen appears, displaying a list of available *Utilities.*

7. If you wish, experiment with the various utilities.

8. To exit Recovery Mode, select the *Apple* menu > *Restart.*

6.2.6   [Apple silicon Mac] Assignment: Boot Into Recovery Mode

*   Prerequisite: Use of an Apple silicon Mac.

1. Power off your Apple silicon Mac computer.

2. Press and hold the power button until you see the *Loading Startup Options* text on the screen.

3. When you see your Mac boot drive and an *Options* button, select the *Options* button, then tap the *Continue* button.

4. At the prompt, select a user account, then tap the *Next* button.

5. At the prompt, enter the user account password, then tap the *Next* button.

6. The Recovery home screen appears, displaying a list of available Utilities.

7. If you wish, you may experiment with the various utilities.

8. To exit Recovery Mode, select the *Apple* menu > *Restart*.

6.2.7   [Intel Mac] Assignment: Boot Into Single-User Mode

In this assignment, you boot into Single-User Mode.

- Prerequisite: Use of an Intel Mac.

1. Power off your Intel Mac computer.

2. Power on your computer, then immediately (before the appearance of the Apple logo) hold down the *cmd* + *S* keys and keep held down.

3. The Login Window appears.

4. Release the cmd + S keys.

5. Select your account, enter your password, then tap the Return/Enter key.

6. The Single-User Mode screen appears, displaying a scrolling list of all commands and activity occurring.

7. The scrolling list of activities stops at a command line prompt.

8. To exit Single-User Mode, enter *exit* at the prompt, then tap the Return/Enter key. The system continues to the normal login window.

6.2.8   [Apple silicon Mac] Assignment: Boot Into Single-User Mode

In this assignment, you boot into Single-User Mode.

- Prerequisite: Use of an Apple silicon Mac.

With Apple silicon Macs and macOS, Apple has removed the previous option of booting into Single-User mode. However, there is still the option to operate from a minimalized System footprint:

1. Power off your Apple silicon Mac computer.

2. Power on your computer, pressing down the Power button and holding it down.

3. Once the *Loading Startup Options* text appears, release the Power button.

4. Select the *Options* button.

5. Once the *Startup Options* screen appears, from the *Utilities* menu, select *Terminal*.

6. The *Terminal* provides a UNIX shell in which to issue commands, without a full Graphical User Interface taking resources.

To exit:

7. Select the *Terminal > Exit Terminal* menu.

8. Select the *Apple* menu > *Shut Down*.

6.3   Resistance to Brute Force Attack

Apple claims there is no back door or golden key to FileVault 2. If there is a way to hack into a FileVault-protected volume, only one group is laying claim to it. *Passware*[5] says their software can break into a FileVault 2 drive in 40 minutes. The author has not tested this claim.

---

[5] *https://www.passware.com/*

6.4   Remotely Access and Reboot a FileVault Drive

When a drive is protected with FileVault, remote support that requires an unattended reboot may become an issue. An example would be a remote server in a locked closet. The reason is that the macOS reboots into an encrypted mode that most remote support software cannot communicate with. So, once the technician reboots the machine, they lose control over it.

A workaround for this situation is to temporarily disable FileVault. This can be done using the Terminal application (located in *Applications > Utilities*) to enter the appropriate command.

6.4.1   [Optional] Assignment: Temporarily Disable FileVault

In this assignment, you temporarily disable FileVault during a macOS restart. This allows remote support software to regain control over the computer after a restart.

- Prerequisite: The computer to be remotely accessed must have the Root user account enabled, and you must know the Root password. See 5.2.1 Assignment: Enable the Root User.

1. While in front of the controlling computer connected to the remote computer (using TeamViewer, ConnectWise, etc.), open *Terminal,* located in *Applications > Utilities* of the remote computer.

2. If you are logged in to the remote computer with a non-administrative user account, in Terminal enter
   `su administrator_name`, and then tap the return key. The su command requests to log in with the specified username.

3. At the Terminal prompt for a password, enter the password for the administrator.

4. Enter the command: `sudo fdesetup authrestart` and then tap the return key. This command requests to disable FileVault on the next restart.

5. At the authentication prompt, enter the administrative password (same as in step 2 above).

6. At the prompt: *Enter a password for '/', or the recovery key*, enter the root password of the local computer.

7. The local computer restarts with FileVault disabled.

8. From the controlling computer viewing the remote computer, at the login screen of the remote computer, enter a *username* and *password* for an account on that computer. You now can work remotely on the computer.

9. On the next reboot, FileVault is enabled as normal.

6.5   Firmware Password

Every device with an operating system has a firmware chip on the logic board. Most people have heard of BIOS, which is the firmware chip on many Windows machines. Firmware has the job of getting things ready for the operating system to take over such as testing major hardware components, verifying voltage levels, managing low-level password security, etc.

Apple computers using an Intel CPU (*not* those using Apple silicon) use a chip on the logic board for part of the boot process. It is the *Extensible Firmware Interface* chip (EFI), normally called the firmware chip. This chip can be password protected. Once password protected, it is not possible to boot the computer from another source (such as Single User Mode, Target Disk Mode, or another startup disk), or use any startup modifier key without first entering the firmware password.

- Note: Apple computers using *Apple silicon* (Apple-designed) CPU do not have the capability or need of a firmware password.

A firmware password on an Intel Mac puts gold plating on your computer security. It is also something to take very seriously. If your Intel Mac was manufactured after January 1, 2010, and you forget your firmware password, the only way to unlock the chip is to physically take the computer to an Apple Store where you may be required to present documents proving your legal ownership of the computer.

6.5.1 [Optional] Assignment: Enable the Firmware Password

In this assignment, install a password on your firmware chip to add the highest-level security to your computer.

- Prerequisite: Intel CPU Mac

1. Boot into Recovery mode with these steps:

   a. Power on your Mac.

   b. Immediately after the startup tone, hold down the cmd + R keys.

2. At the *macOS Recovery* screen, select the boot volume. Then tap the *Next* button.

3. Select a *User*, then tap the *Next* button.

4. Enter the User's password, then tap the *Continue* button.

5. Select the *Utilities* menu > *Startup Security Utility*.

6. At the *Startup Security Utility* window, tap the *Turn On Firmware Password* button.

7. Enter a strong password, confirm it, then tap the *Set Password* button.

8. Record your password in a secure location.

9. Tap the *Quit Startup Security Utility* button. The step returns you to the main Recovery HD screen.

10. Restart your Mac by selecting *Apple* menu > *Restart*.

Congratulations!

6.5.2 [Optional] [Intel Mac and Windows] Assignment: Test the Firmware Password

In this assignment, you verify that your firmware password is active.

1. Shut down your Mac.

2. Power on your Mac. Immediately after the startup tone, hold down the option key.

Without a firmware password, this startup modifier would put you into *Start Manager Mode*, allowing anyone full access to all your data by booting from an external drive. But with a firmware password in place, you should see a screen requesting the firmware password to proceed.

3. Enter your firmware password.

4. Select your normal boot volume then continue startup as normal.

6.5.3 [Optional] [Intel Mac] Assignment: Remove the Firmware Password

I consider having a firmware password in place as important as the user password. The times when removing the Firmware Password is called for include:

*   Selling or giving away the computer

*   Bringing in the computer for hardware or software service

In this assignment, you remove the Firmware Password. If you wish to leave it enabled, skip this assignment.

1. Boot into Recovery HD mode:

    a. Power on your Mac.

    b. Immediately after the startup tone, hold down the cmd + R keys.

2. At the *macOS Recovery* screen, select the boot volume, then tap the *Next* button.

3. Select a *User*, then tap the *Next* button.

4. Enter the User's password, then tap the *Continue* button.

5. Select the *Utilities* menu > *Startup Security Utility*.

6. At the Startup Security Utility window, tap the *Turn Off Firmware Password* button.

7. Enter the Firmware Password, then tap the *Turn Off Firmware* button.

8. Tap the *Quit Startup Security Utility* button. This action returns you to the main Recovery HD screen.

9. Restart your Mac by selecting *Apple* menu > *Restart*.

6.5.4 Assignment: Startup Security for Apple silicon Mac

Macintosh computers that use the new Apple silicon CPU instead of Intel chips do not have an EFI chip. However, they do have military-grade startup security built into the Apple silicon. You have control over this level of security.

In this assignment, you boot into Recovery mode to configure your Startup Security.

- Prerequisite: A Mac with an Apple silicon CPU.

1. If powered on, shut down your Mac.

2. Power on by pressing and holding down the power button. Hold down the power button until you see *Loading Startup Options*

3. Select the *Options* icon, then tap *Continue.*

4. At the *MacOS Recovery* screen, select an administrator account, then tap *Next.*

5. Enter the administrator password, then tap *Continue.*

6. Select *Utilities* menu > *Startup Disk Utility.*

7. Select the boot volume, then tap *Unlock.*

8. Select a username, enter their corresponding password, then tap *Unlock.*

9. Select a volume, then tap the *Security Policy* button. The *Startup Security Utility* window opens.

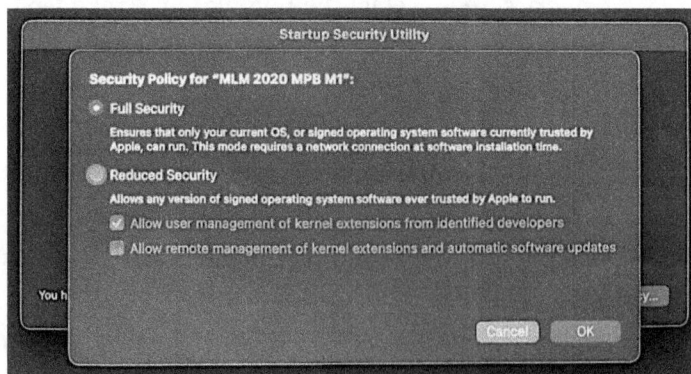

10. For the highest level of security, select *Full Security*. This level requires that this computer can only be booted by an Apple-signed OS of current vintage. This is important because older versions may have vulnerabilities, and if used to boot your Mac, older versions could make your data accessible.

11. Tap *OK*.

12. If prompted, enter an administrator username and password.

13. Select *Apple menu > Restart* or *Shut Down* to enable the new Startup Security Policy.

6.5.5   [Windows] Assignment: Enable Additional UEFI/BIOS Security Features

6.6   [Android and iOS] SIM Card Lock

6.6.1   [Android and iOS] Assignment: Set Up a SIM Card Lock

6.7   [iOS] Smart Card Lock

6.7.1   [iOS] Assignment: Set Up A SIM Card Lock

6.8   Camera and Microphone Recording Indicator and Hardware Disconnect

Camera and Microphone Indicator

New with macOS 12, iOS 14 and iPadOS 14 is a menu bar indicator whenever an app is accessing the device's microphone or camera. An orange dot indicates microphone access, a green dot indicates camera access. Tapping on the indicator displays which app(s) are accessing the camera or microphone.

6.8.1   Assignment: See the Recording Indicator in Action

In this assignment, you will see the recording indicator in action.

Test Microphone Indicator

1. Look to the right side of your menu bar. You should not see any small colored dot.

2. Open *QuickTime Player.* You should still not see any small colored dot.

3. From the QuickTime Play app menu, select *File > New Audio Recording.*

4. Look again at the right side of the menu bar. You will now see a small orange or red dot (in this screenshot, located just above the "A" in "Audio Recording." This is because you have instructed an app to access the microphone:

5. Tap on the small colored dot. A pop-up menu will display the apps accessing the microphone at the very top of the menu:

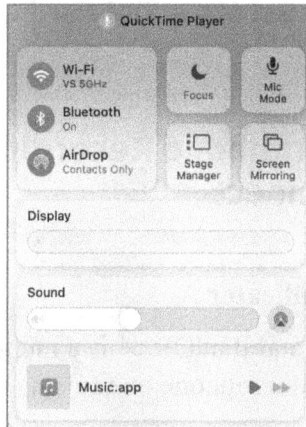

6. Tap on the desktop to make the menu go away.

Microphone Hardware Disconnect

All Macs with Apple silicon or the Apple T2 security chip feature a microphone hardware disconnect[6] when the lid is closed. This fully prevents any software from accessing the microphone with a closed lid.

6.8.2   [Optional] Assignment: Test Microphone Hardware Disconnect

In this assignment, you verify the microphone hardware disconnect works.

- Prerequisite: A Mac laptop with built-in microphone.

- Prerequisite: A Mac with either Apple silicon or an Apple T2 security chip[7].

7. Open the lid on your Mac laptop.

8. Open *System Settings > Sound > Input* tab.

9. Select the built-in microphone:

---

[6] *https://support.apple.com/guide/security/hardware-microphone-disconnect-secbbd20b00b/web*

[7] *https://support.apple.com/en-us/HT208862*

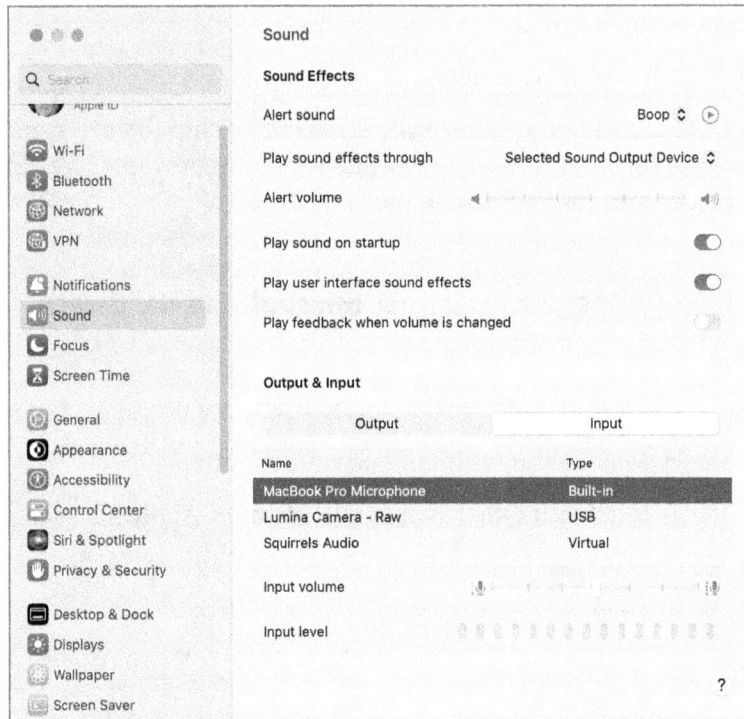

10. Close *System Settings.*

11. Open *Quick Time Player.*

12. Select the *File* menu > *New Audio Recording,* then tap the *Record* button.

13. Speak a few words, then continue speaking while closing the lid, and continue speaking while opening the lid.

14. Stop the recording.

15. Play the recording. Note how when the lid closed, recording stopped, recording a blank space.

16. Quit QuickTime Player without saving the file.

6.9   Siri Offline Command Processing

As of macOS 12 and iOS 15 (for iOS and iPadOS devices with an A12 Bionic processor or later) is Siri offline command processing.

Prior to this, all Siri commands were sent via internet to Apple servers for processing–even for commands destined for local implementation, such as *Open camera*.

Now most commands will be processed on the local device. Only commands requiring external resources–such as weather forecasts–will submit the verbal command to Apple's servers.

This goes a very long way to ensuring your privacy.

6.10   Device Hardware Lessons Learned

☐ macOS 13 does not have built-in tools to block access to physical ports. Third-party End Point Protection software is required if you must block access to ports.

☐ FileVault 2 Full Disk Encryption is built into macOS.

☐ Target Disk Mode allows almost any Mac to act as an external storage device.

☐ Recovery Mode allows booting into the otherwise invisible Recovery partition of the boot drive to perform directory repair, reinstall macOS, enable Firmware password (Intel Mac only), etc.

☐ Single-User Mode allows booting into a command-line state with only the root user account active.

☐ Boot into Target Disk Mode on an Intel Mac by holding down the T key immediately after power on, then releasing after the login window appears.

☐ Boot into Target Disk Mode on an Apple Silicon Mac by continuing to hold the power button after power on until *Loading startup options* appears. Then select *Options* button > *Continue* button > *Utilities* menu > *Share Disk*.

☐ Boot into Recovery Mode on an Intel Mac by holding down *cmd + R* immediately after power on, releasing after the login window appears.

☐ Boot into Recovery Mode on an Apple Silicon Mac by holding down the power button after power on, then releasing when you see the *Loading Startup Options* text > *Options* > *Continue* > *Next* > *Next*.

☐ Boot into Single-User Mode on an Intel Mac by holding down *cmd + S* immediately after power on, releasing when the login window appears.

☐ Boot into Single-User Mode on an Apple silicon Mac by holding down the power button after power on, then releasing when the *Loading Startup Options* appears. Select *Options* > *Utilities* menu > *Terminal*.

☐ To enable FileVault on a non-boot volume, *right-tap* on the drive > *Encrypt*.

☐ The firmware (EFI) chip on an Intel Mac can be hardened with its own password.

☐ To enable the firmware password on an Intel Mac, boot into Recovery Mode (*cmd + R*) > at the *macOS Recovery* screen. Then select the boot volume > *Next* > select a *User* > *Next* > enter the *User Password* > *Continue* > *Utilities* menu > *Startup Security Utility* > *Turn On Firmware Password*.

☐ Instead of a firmware password, Apple silicon Macs offer a *Startup Security* option.

☐ Access *Startup Security* on an Apple silicon Mac by holding down the power button after power on, then releasing when *Loading Startup Options* appears > *Options* > *Continue*. At the *macOS Recovery* screen, select an administrator account > *Next* > enter a *Password* > *Continue* > *Utilities* menu > *Startup Disk Utility*. Select the boot volume > *Unlock*. Select a *Username* > enter a *Password* > *Unlock*. Select a volume > tap *Security Policy* button, then select your desired level of startup security.

☐ New as of macOS 12 is microphone hardware disconnect, preventing any software from accessing the microphone when the lid of a portable Mac is down.

☐ New as of macOS 12 is camera and microphone recording indicator. When an app accesses either the camera or microphone, a small colored dot appears near the right side of the menu bar. Tapping on the dot displays the apps in use.

☐ New as of macOS 12 is that most Siri commands are now processed locally, without having to be sent to Apple's servers.

7 Sleep and Screen Saver

Do not take life too seriously. You will never get out of it alive.

–Elbert Hubbard[1], American writer, publisher, artist, and philosopher

What You Will Learn in This Chapter

- Configure password after Sleep or Screen Saver
- Configure Lock Screen Notifications
- Enable Do Not Disturb

What You Will Need in This Chapter

- No additional resources required.

7.1 Require Password After Sleep or Screen Saver

When you are not actively using your powered-on computer, by default it remains on even if the display goes to sleep. It is easy for someone else to sit in front of the computer and access all your data.

To help prevent this access, configure your computer to lock down after a short period of inactivity (5-15 minutes), or upon command.

---

[1] *https://en.wikipedia.org/wiki/Elbert_Hubbard*

7.1.1 Assignment: Configure Screen Saver

In this assignment, you configure the computer to require entering a password to remove the screen saver if more than 5 seconds after the screen saver launches, and to go into screen saver mode after 5 minutes of inactivity.

Configure Require Password 5 Seconds After Sleep or Screen Saver Begins

1. Open *Apple* menu > *System Settings* > *Lock Screen*. Configure to your taste. Shown below are my recommended settings:

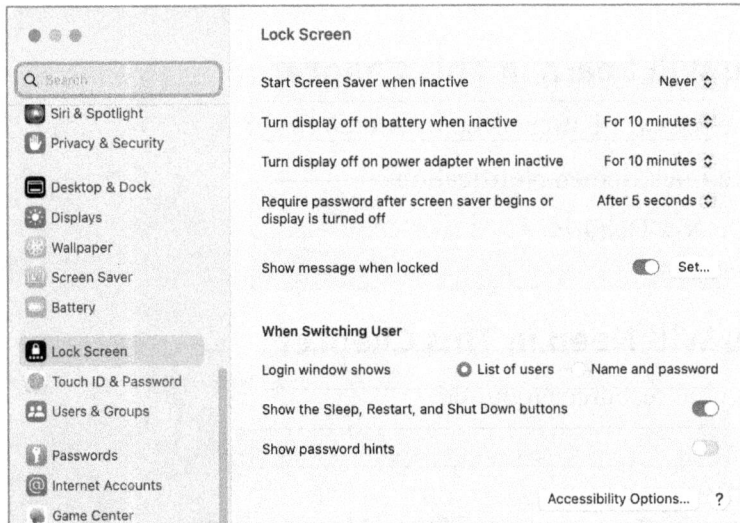

Configure Hot Corner for Sleep

2. From the *System Settings* sidebar, select *Desktop & Dock* > scroll to the bottom > tap *Hot Corners*.

3. Select one of the four corners pop-up menus, then select *Start Screen Saver*. This puts your display into sleep mode the moment your cursor is pushed to that corner. Then select the *OK* button.

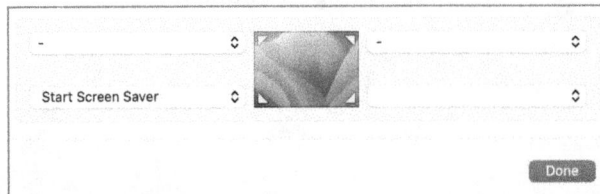

4. Quit System Settings.

5. Test your new setting by pushing your cursor into that corner. The display should go into sleep mode. Leave the cursor there for 5 or more seconds, then move the cursor. You should be presented with an authentication screen.

6. Enter your password to unlock your computer and return to work.

You have successfully enabled lockdown of your system with sleep or screensaver modes.

7.2   Lock Screen Notifications

If you've ever had an embarrassing text message show up on your device while meeting with business clients or friends, you know the importance of restricting lock screen notifications. In addition, for some applications this can pose a serious security risk as these notifications on your lock screen may be viewed by anyone.

7.2.1   Assignment: Restrict Lock Screen Notifications

In this assignment, you configure which notifications may display while your device is in lock screen mode.

1. Open *Apple* menu > *System Settings* > *Notifications*. The main area lists all applications installed that have a notification feature.

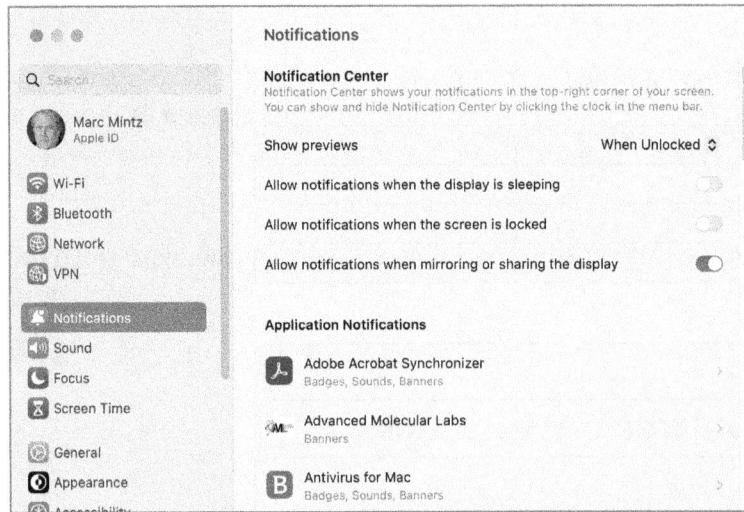

2. For best privacy, disable *Allow notifications when the display is sleeping,* and *Allow notifications when the screen is locked.*

3. Starting with the first application listed in the main area, select the application, then configure how notifications should appear to your taste. When done, repeat for the next application.

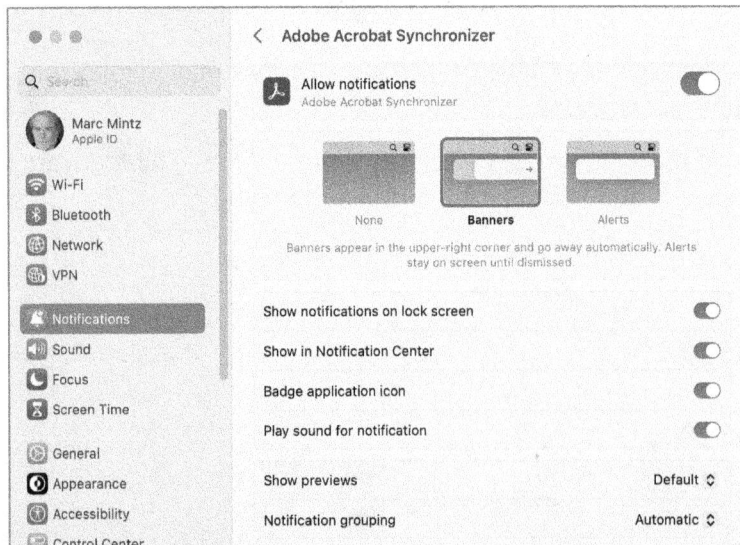

7.3   Do Not Disturb Mode

If you ever had your phone go off during an important meeting (or 3am), then you know the value of having *Do Not Disturb mode* in place. This mode allows for you to set times and days when you would like *Do Not Disturb* to be on, as well as what can come through as an exception.

7.3.1   Assignment: Configure Do Not Disturb Mode

In this assignment, you enable *Do Not Disturb*.

1. Select *Apple* menu > *System Settings* > *Focus*.

2. Tap the *Add Focus...* button.

3. In the *What do you want to focus on?* window, select *Reading*. The *Reading* Focus configuration window appears.

4. From the sidebar, tap *Do Not Disturb*.

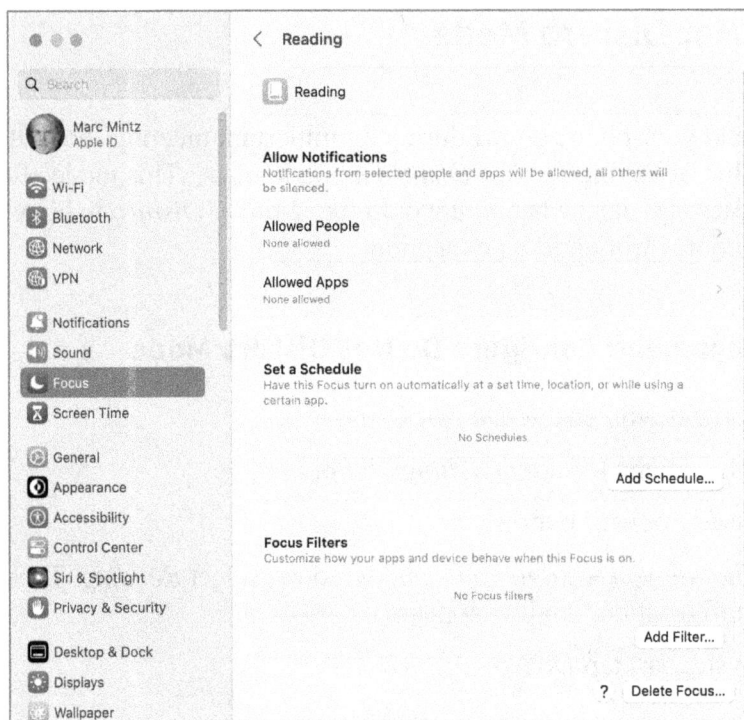

5. In the *Allowed Notifications* area, tap *Allowed People*. Configure to your taste to allow or prevent some people to send you notifications or calls. When done, tap *Done*.

6. When done, close *System Settings*.

Activate a Focus Profile

7. Open *System Settings > Control Center > Focus*.

8. Set *Focus* to *Always Show in Menu Bar*. *Focus* icon will display in the menu bar.

9. Tap the *Focus* icon in the menu bar > *Reading*. The *Do Not Disturb* settings for *Reading* will be in effect until you turn off the *Reading* Focus.

Deactivate a Focus Profile

10. Tap the *Focus* icon in the menu bar > *Do Not Disturb*. The *Focus* icon will turn gray/off.

7.4 Sleep and Screen Saver Lessons Learned

☐ It is important to require a password after waking from sleep or screen saver so that passers-by cannot access your device if you are away. Configuration is accessed from *Apple* menu > *System Settings* > *Lock Screen*.

☐ Restrict Lock Screen Notifications from *Apple* menu > *System Settings* > *Focus*.

☐ Configure a Hot Corner to immediately put the screen to sleep from *System Settings* > *Desktop & Dock* > scroll to the bottom > tap *Hot Corners*.

8 Malware

Behind every great fortune lies a great crime.

–Honore de Balzac[1], 19th-century novelist and playwright

What You Will Learn in This Chapter

- Understand anti-malware
- Select an anti-malware utility
- Install and configure Bitdefender
- Scan for malware

What You Will Need in This Chapter

- No additional resources required.

8.1   Anti-Malware

Most people know this category of software as Antivirus. But there are so many other nasty critters out there (worms, Trojan horses, phishing attacks, malicious scripts, spyware, etc.) that the overarching term "Anti-Malware" is more accurate.

Depending on how one chooses to measure, there are from 500,000–40,000,000 malware[2] in the field that impact Android, Chrome OS, iOS, macOS, Windows,

---

[1] *https://en.wikipedia.org/wiki/Honoré_de_Balzac*
[2] *http://en.wikipedia.org/wiki/Malware*

and the Internet of Things (IoT). Symantec reports they receive as many as 40,000 new signatures in a single day.

macOS includes several automatically updating system-level architectures designed to help prevent malware from getting a foothold. Although Apple has done a good job with anti-malware, they can do better. These invisible utilities only protect against known macOS malware, and Apple has been slow to update when new malware shows up. Should we care about Windows malware? Yes. Windows-specific malware may not hurt you unless you have a virtual machine running Windows. But it is probable that at some point you will inadvertently pass along Windows malware to a friend or business associate who is using Windows. Imagine how your relationship will change should an email from you take down a friend's computer or a customer's entire network.

Should you and I care about malware? Yes. Not only can and will malware slow the computer and cause data and backup corruption (usually an unintended malware consequence), but your every keystroke, passwords, email, and web browsing may be harvested. This information is commonly sold on hacker sites, allowing anyone with a few dollars to steal your identity, bank account, credit cards, etc.

For these reasons, I strongly recommend you install quality anti-malware on all macOS machines. This raises the question: how to know an anti-malware is quality software? We go by the results of independent testing organizations. These include AV TEST[3], ICSA Labs[4], and AV Comparatives[5] (AVC). Although no testing organization tests all 100+ anti-malware products on the market, AVC tests the major players at least a few times each year against a wide range of current malware. The results of AVC's Android, macOS, and Windows anti-malware product tests are made public on the AVC website[6, 7].

In their most recent testing of macOS anti-malware products, most tested software caught all the OS X malware. So, the deciding factors come down to ease of use, resource utilization (impact on computer performance), and ability to catch Windows malware.

---

[3] *https://www.av-test.org/en*

[4] *https://www.icsalabs.com/technology-program/anti-virus*

[5] *http://av-comparatives.org*

[6] *http://www.av-comparatives.org/comparatives-reviews/*

[7] *http://www.av-comparatives.org/mac-security-reviews/*

The only product we currently recommend for home and business users is Bitdefender[8]. This is due to their first-rate ability to recognize and remove malware of all platforms, simple interface, low impact on computer performance, and offering of Android, Chrome OS, macOS, and Windows versions. If you have more than a few computers, you can upgrade to Bitdefender's business version which is centrally administered from a cloud-based console. This option is called *GravityZone*.

For macOS users running Windows in a Virtual Machine environment, you also need an anti-malware utility running on Windows. Although Microsoft provides a free option, by most independent testing reports it can catch only 95% of known malware. This is about the same as leaving for vacation after locking the front door but leaving the back door open.

Here is AVC's chart on the effectiveness of catching and removing malware from Windows:

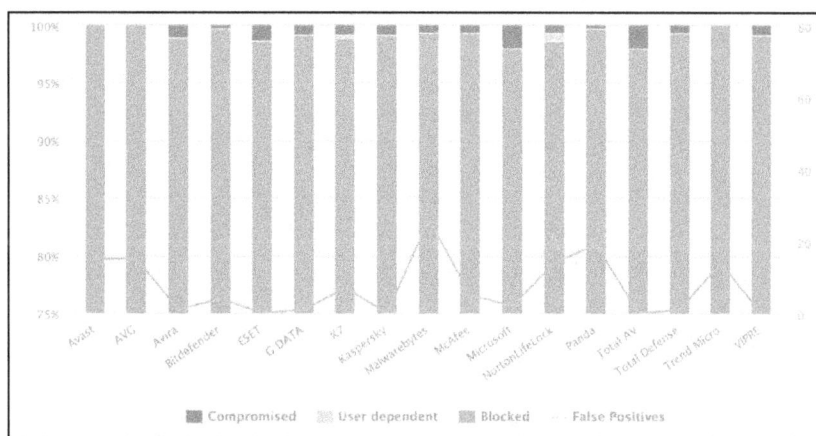

Effectiveness isn't the only important measurement. In this same chart, we see false positives. It's a sad day when a clean file is flagged as infected then automatically trashed.

Performance is another vital measurement. This is the impact the anti-malware has on overall performance of your computer. Anti-malware should use as little system resources as possible. Compare the performance shown below:

---

[8] *http://www.bitdefender.com/*

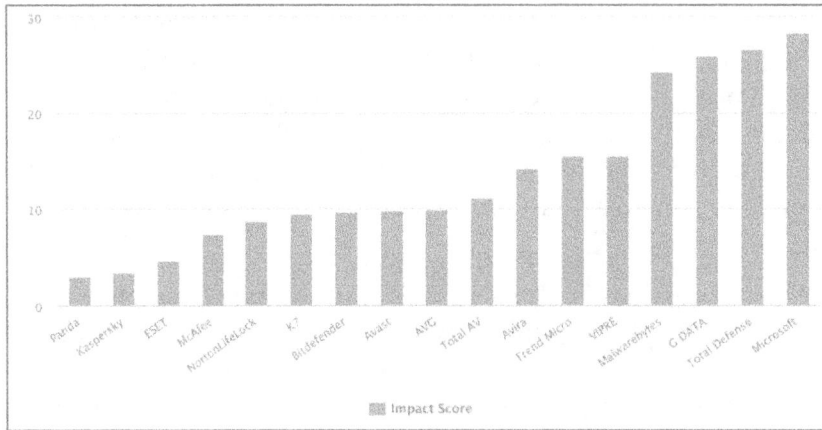

AV-Comparatives performs separate testing for Android and macOS anti-malware products[9]. They do not conduct as thorough testing for these products as for Windows. A summary of product testing is provided by AV-Comparatives below.

---

[9] *https://www.av-comparatives.org/wp-content/uploads/2017/08/avc_mac_2017_en.pdf*

---

8 Malware

Mobile Protection Rates		
	Protection Rate	False Positives
Bitdefender, Trend Micro	100%	0
Avira, G DATA, Kaspersky	99.9%	0
Avast	99.9%	1
AVG	99.9%	2
Securion	99.6%	2
Google	91.3%	7

Vendor	Test	Award	Platform
Avast Security for Mac	Mac Security Test & Review 2020	APPROVED	MacOS
AVG Internet Security for Mac	Mac Security Test & Review 2020	APPROVED	MacOS
Avira Antivirus Pro for Mac	Mac Security Test & Review 2020	APPROVED	MacOS
Bitdefender Antivirus for Mac	Mac Security Test & Review 2020	APPROVED	MacOS
CrowdStrike Falcon Prevent for Mac	Mac Security Test & Review 2020	APPROVED	MacOS
FireEye Endpoint Security for Mac	Mac Security Test & Review 2020	APPROVED	MacOS
Kaspersky Internet Security for Mac	Mac Security Test & Review 2020	APPROVED	MacOS
Trend Micro Antivirus for Mac	Mac Security Test & Review 2020	APPROVED	MacOS
Pocket Bits BitMedic Pro Antivirus	Mac Security Test & Review 2020	NOT APPROVED	MacOS

As with any research, do not use just one set of data points. The "winners" and "losers" jostle for position monthly.

8.1.1 Assignment: Install and Configure Bitdefender for Home Users

In this assignment, you download, install, and configure Bitdefender Antivirus for Mac.

- Note: Bitdefender Antivirus for Mac is for home users only. If you are a business, skip this assignment, and perform the next assignment.

1. Using your favorite browser, visit *https://Bitdefender.com/Downloads.*
2. Select *Download Now* button.
3. Locate the installer in your Downloads folder. Then double tap to launch the installer. It mounts and opens a virtual disk on your Desktop.
4. Follow the on-screen instructions to install Bitdefender downloader.
5. Once installed, the Bitdefender Antivirus for Mac downloads to your computer.
6. Once the download completes, installation begins automatically.
7. As a security precaution, macOS automatically blocks programs that request full disk access—such as an antivirus application. To install:

   a. At the *Grant Bitdefender Full Disk Access* window, tap the *Open System Settings* button.

176

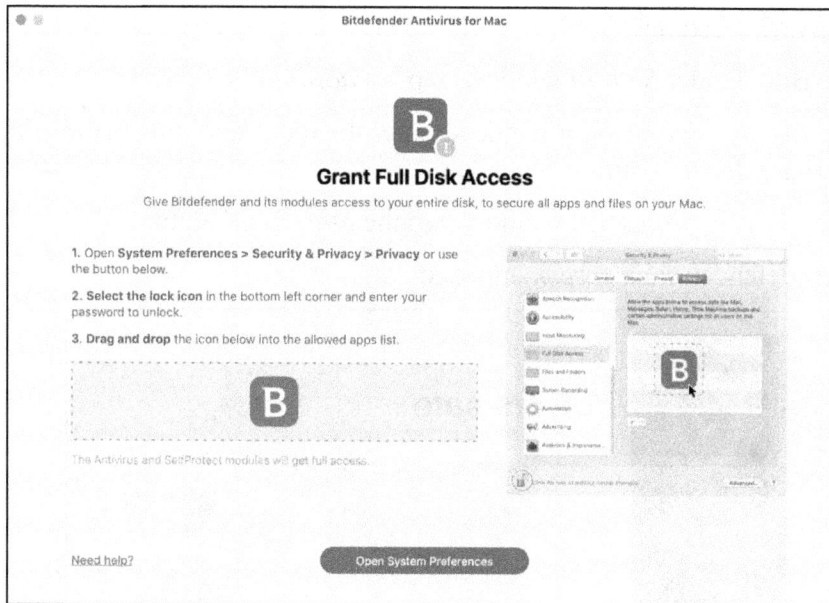

b. Unlock the *Lock* icon.

c. Select *Full Disk Access* from the main area.

d. Drag the Bitdefender B icon from *Grant Full disk Access* window of the installer into the right side of *Privacy* window.

e. Enable all *Bitdefender* or *Antivirus for Mac* checkboxes.

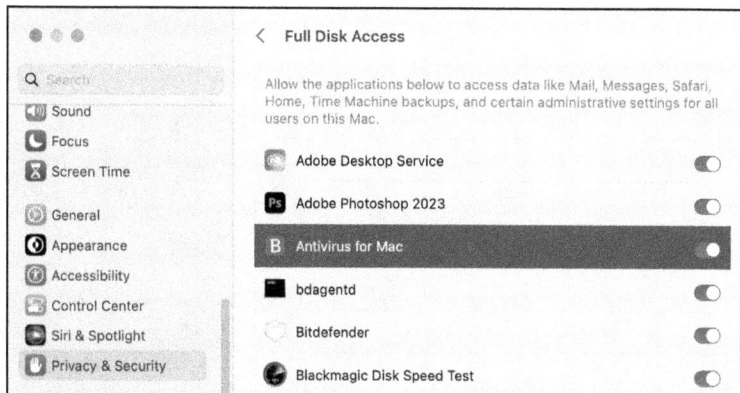

8. At the prompt, select *Quit and Restart Bitdefender*.

9. In the Bitdefender *login is required* window, tap the *Login* button.

10. Quit System Settings.

11. In the Bitdefender *Sign In* window, tap *No account? Create one*.

12. In the *Create your account* window, enter the requested information. Then tap the *Create Account* button.

13. At the *Activate protection* window, tap the *Start Trial* button.

14. The Bitdefender *main screen* appears:

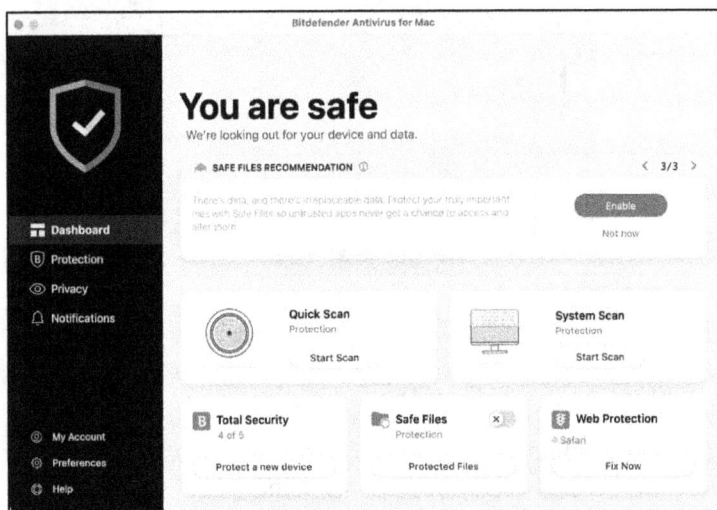

15. At the *Allow security modules to run* window, tap *Security & Privacy* link:

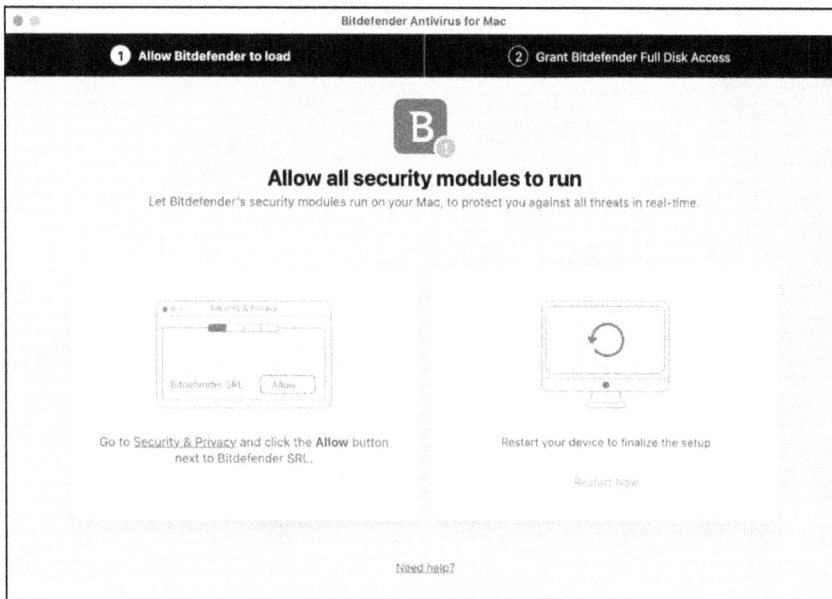

16. Follow the on-screen instructions to permit the Bitdefender extension to run on your system.

17. Restart your computer to allow Bitdefender to finalize setup.

18. After your computer has restarted, open Bitdefender.

19. Select *Bitdefender* menu > *Preferences*. The Preferences window opens.

20. Select the *Protection* tab. Configure to your taste. My settings are shown below:

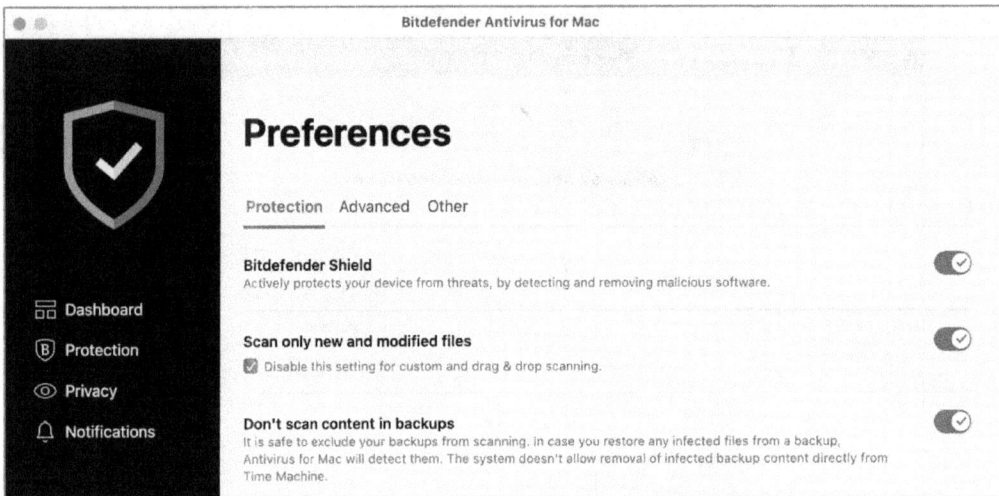

21. Tap the *Advanced* tab. Configure to your taste. My settings are shown below:

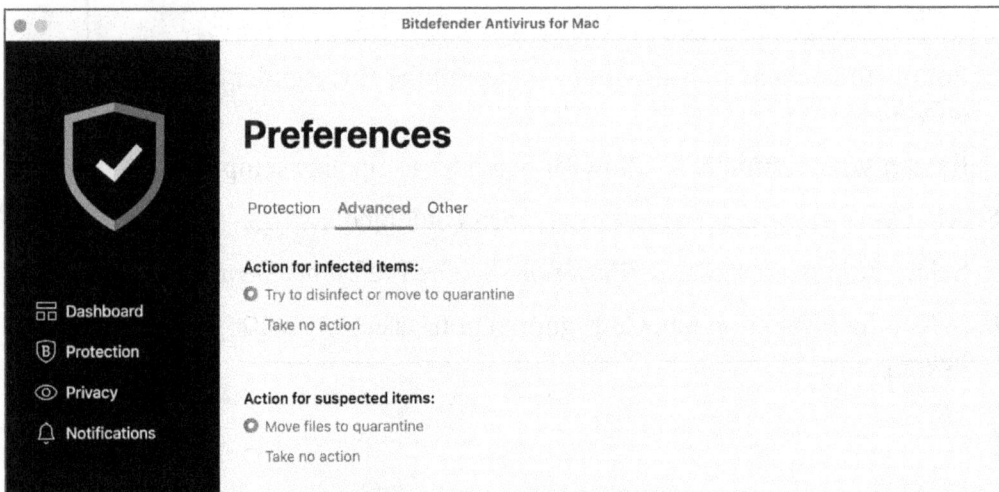

22. Tap the *Other* tab. Configure to your taste. My settings are shown below:

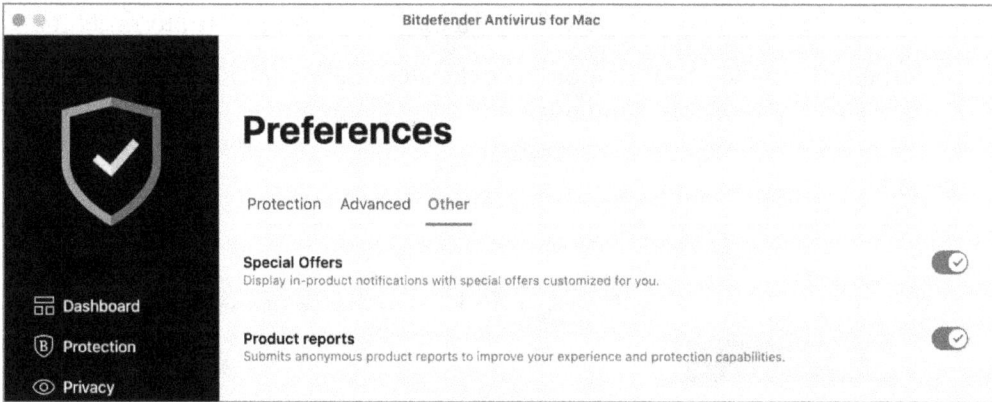

23. From the side bar, tap *Dashboard.*

24. To ensure your system is free from malware, perform an initial full system scan. In the *System Scan Recommendation,* tap the *Scan* button.

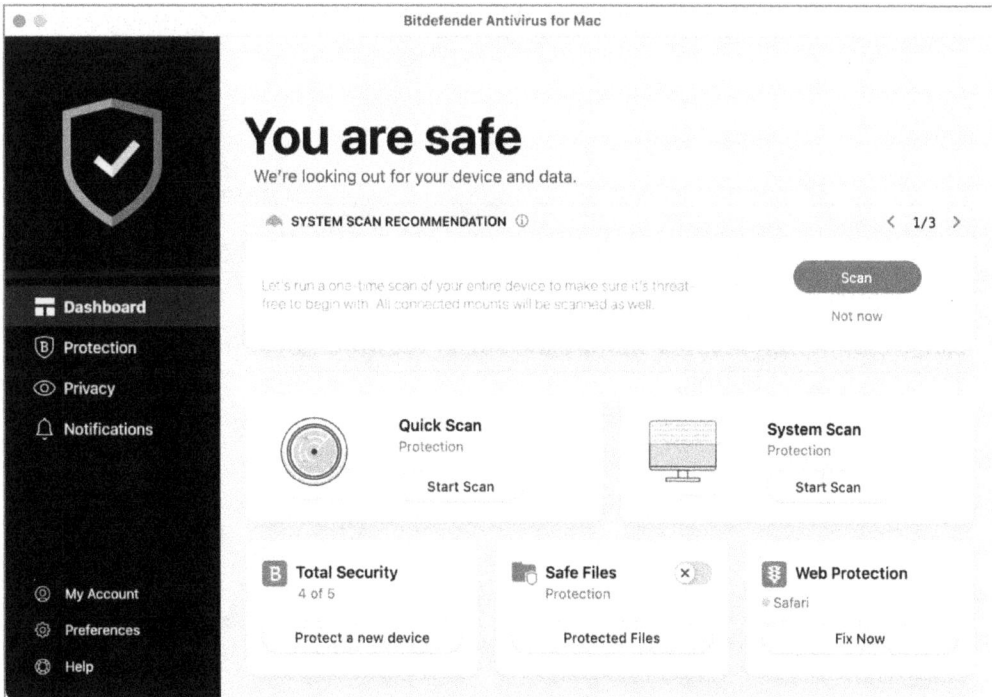

25. While the scan process continues (it may take several hours depending on the speed of your system and number of files), return to the main Bitdefender window, and from the *TrafficLight Recommendation* area, tap *Install* button.

- Note: You may need to cycle through the *Notifications* area to find this.

26. Bitdefender starts to install the Safari extension. Safari opens to the *Preferences > Extensions* window.

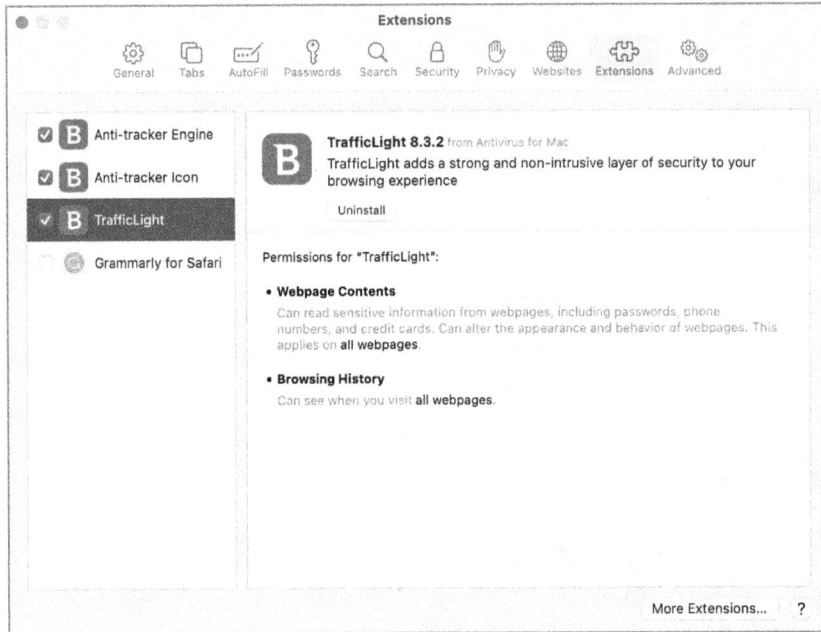

27. Enable the checkboxes for the three Bitdefender Trafficlight extensions, then quit Safari.

28. Return to the Bitdefender main window. In the *Safe Files Recommendation* area, tap *Enable* button.

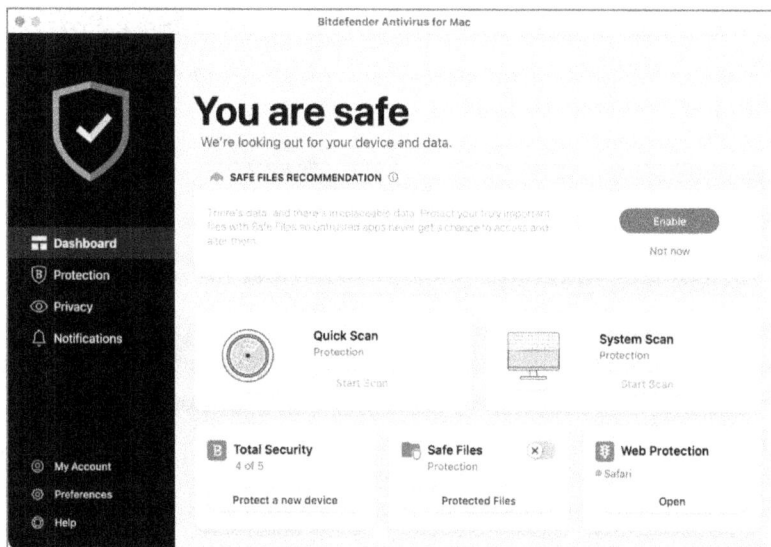

29. Allow the Bitdefender *Full System Scan* to continue before quitting Bitdefender.

8.1.2   Assignment: Install and Configure Bitdefender for Business Users

In this assignment, you download, install, and configure *Bitdefender GravityZone Endpoint Security.* The advantages of using a business-class product instead of a consumer-class AV product include:

- Unlimited number of devices under one account (based on pricing)
- More granular control over protection settings
- Server protection
- Mobile device protection
- Microsoft Exchange protection
- Note: Bitdefender GravityZone is for business use. If you are installing for home use, skip this, and perform the previous assignment.

Install Bitdefender GravityZone

1. Go to *https://bitdefender.com.*

2. From the menu, select For *Business* > *Products* > *Compare Products*.

3. Review all product options to determine which would be the best fit for your organization.

- Note: For this assignment, you use *GravityZone Elite.*

4. Under the *GravityZone Elite* column, select *Try for Free.*

5. Follow the onscreen instructions to download and install GravityZone Elite.

6. Configuring GravityZone Elite is significantly more complex than the home versions. Once you have installed GravityZone onto one of your computers, visit the Bitdefender support page for documentation on configuration.

Congratulations! You have just built a wide moat to keep known malware from your computer.

8.1.3 [Windows] Assignment: Disinfect Windows From Outside Windows

8.2 Remove Malicious or Problem Applications

Malicious applications come from a variety of sources:

- Application designed to be malicious. Called *Trojan Horses,* these applications are designed to look like something desirable such as a game but may be harvesting your data in the background.

- Infected application. Applications may have been intentionally or unintentionally infected at the developer, the duplication center, the distribution center, or on your device.

- Infected file. Malware is often designed to infect files. If you send an infected file to another computer, the infection spreads to other files on that computer.

Although Apple does a decent job of weeding out malicious and problem apps from the App Store, some still get through. In addition, some otherwise quality apps may not play well on your device due to incompatibility issues. If you have downloaded apps from other than the App Store, it is likely they have not been vetted, so all bets are off on them.

If you find that your device is having problems, such as:

- Pop-up ads or tabs that refuse to close or go away

- Home page changes

- Search engine changes

- Your web browsing gets redirected

- Virus alerts from other than your installed anti-virus software

Your problem may be due to an infected, malicious, or problem app.

8.2.1 Assignment: Isolate and Remove Malicious or Problem App with Bitdefender

In this assignment, you find then remove a problem application.

1. After Bitdefender is installed, open Bitdefender.

2. Select *Actions* menu > *Update Threat Information Database.* Bitdefender automatically updates itself when connected to the Internet, but this manual check verifies you have the latest anti-malware database installed.

3. From the main Bitdefender window, tap *Quick Scan.* This will ignore any files previously scanned and scan only newer files.

4. Once complete, a list of any infected items displays.

5. Most malware can be automatically removed by Bitdefender, leaving the original application or file intact. If this cannot be done, Bitdefender alerts you, giving the option to move the item to quarantine (a special folder for infected items) or to delete the item. Generally, it is best to delete the item, unless it is a file that is the only source of vital data. In that case, copy the file to an offline computer dedicated to only opening suspicious files, open the file, then copy the data to a newly created clean file.

6. Shut down the computer. Do this because there is a minor history of malware getting into memory. A shut down purges whatever may be in memory.

7. Wait a minute, then power on.

8. Get back to work on your computer.

Technically, you should never have to do a System Scan after the first time, because Bitdefender scans all items as they open. However, I recommend performing a monthly System Scan to catch any infected items that entered your computer and were scanned before their threat information was added to the database.

8.2.2   Assignment: Protect Your Device from Problem Apps

Apple has made it extremely difficult for malicious applications to be installed on your machine. Built into the OS is a module called *Gatekeeper*. It whitelists which applications can run on the computer based on whether they downloaded from the *Apple App Store* or from *Apple App Store and identified developers*. If an app was downloaded from anywhere else, it probably cannot run.

Because of this, it is recommended to *only* download software from either:

• Apple App Store

• Directly from the developer's website

In this assignment, you configure Gatekeeper to allow apps from the Apple App Store and identified developers only.

1. Open *System Settings > Privacy & Security*.

2. Scroll down the main area to the *Security* section.

3. Select *App Store and identified developers*.

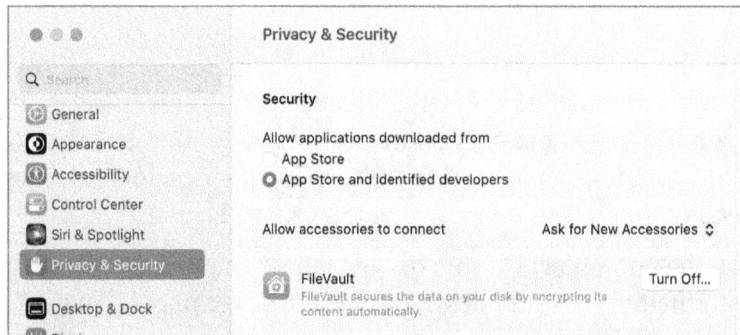

4. Exit *System Settings*.

For most of us, configuring Gatekeeper to allow *App Store and identified developers* is a reasonable security setting. Some highly restrictive environments may wish to limit to *App Store* only.

8.2.3 [Android, Chrome OS, macOS, and Windows] Assignment: Protect Your Device from Problem Apps with Bitdefender

8.3 [Windows] Controlled Folder Access

8.3.1 [Windows] Assignment: Enable Controlled Folder Access

8.4 Advanced Outlook.com Security for Office 365 Subscribers

When using Outlook.com as an Office 365 subscriber, an advanced security module called *Safelinks*[10] is active by default. Safelinks performs extra screening of attachments and links in messages you receive.

Attachments

As a paid subscriber, Safelinks scans all attachments for viruses and malware. If it detects a suspicious file, it is removed so you can't accidentally open it.

Links

When receiving a message with a web link, Safelinks scans to see if the link is related to phishing scams or might download malware. If you do tap a link that is suspicious, you are redirected to a warning page:

---

[10] *https://support.microsoft.com/en-us/office/advanced-outlook-com-security-for-office-365-subscribers-882d2243-eab9-4545-a58a-b36fee4a46e2*

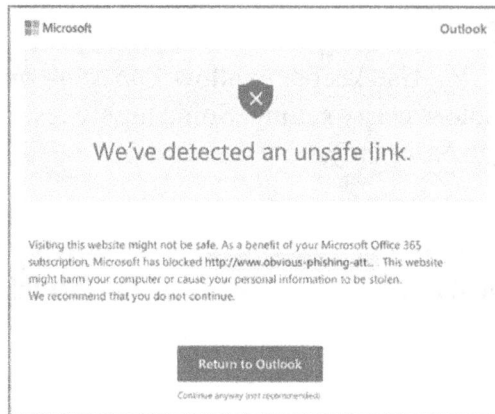

Safelinks Protection

This protection is automatically active with a paid 365 subscription, regardless of how you access your email–from outlook.com or any email client. The protection is only for Outlook.com mailboxes only, and do not apply to third-party accounts such as Gmail that are synched to an Outlook.com account.

8.4.1   [Optional] Assignment: Disable Safelinks in Outlook.com

In this assignment, you disable Safelinks in Outlook.com email.

- Note: This assignment is provided only for informational purposes. We do not recommend disabling Safelinks.

- Prerequisite: A paid Office 365 account with an active Outlook.com email account.

1. Open a browser to *https://outlook.live.com.*

2. Select *Settings > Premium > Security*.

3. Under *Advanced Security,* turn off the switch for *Safelinks*.

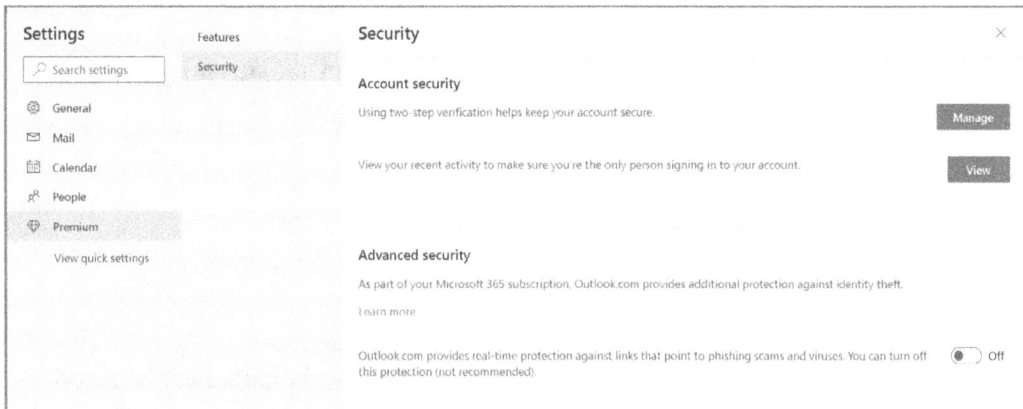

- Note: Turning off Safelinks only affects future messages. It does not change the link format in messages already received.

8.5   [Windows] Windows Security Configuration

8.5.1   [Windows] Assignment: Harden Windows Security Configuration

8.6   Lockdown Mode

New to iOS 16 and macOS 13 is *Lockdown Mode*. This brings extreme cybersecurity to iPhone/iPad and Mac users. Although aimed at the small minority of users that may be the subject of highly targeted spyware, we hold that all users should evaluate if such heightened security is appropriate for them.

Once enabled, Lockdown Mode adds the following protections, with Apple promising to continually expand:

- Apple Services
  - o Incoming invitations for Apple services from people you haven't previously invited are blocked.

- Browsers using WebKit (including Safari)

  o Some web technologies such as just-in-time (JIT) JavaScript are disabled until the user excludes a trusted site from Lockdown Mode.

- Configuration profiles

  o Profiles cannot be installed.

  o Device cannot enroll into mobile device management (MDM).

- FaceTime

  o Incoming calls from people you have not previously called are blocked.

- Hardware

  o Wired connections with other devices and accessories are blocked when the device locks.

- Messages app

  o Message attachments other than images are blocked.

  o Link previews do not display.

- Photos app

  o Shared albums are removed.

  o New shared album invitations are blocked.

8.6.1   [Optional] Assignment: Enable Lockdown Mode

In this assignment, you enable Lockdown Mode.

1. Open *System Settings > Privacy & Security > Lockdown Mode.*

2. Tap *Turn On.* The *Lockdown Mode* screen appears.

3. Tap *Turn On Lockdown Mode.*

4. Authenticate, then tap *Modify Settings.*

5. In the *Turn on Lockdown Mode?* window, tap *Turn On & Restart.*

6. Tap *Turn On & Restart.*

7. Your device restarts with Lockdown Mode enabled.

8.6.2 [Optional] Assignment: Disable Lockdown Mode

In this assignment, you disable Lockdown Mode.

1. Open *System Settings > Privacy & Security > Lockdown Mode.*

2. Tap *Turn Off.* The *Lockdown Mode* screen appears.

3. Tap *Turn Off Lockdown Mode.*

4. Authenticate, then tap *Modify Settings.*

5. In the *Turn off Lockdown Mode?* window, tap *Turn Off & Restart.*

6. Tap *Turn Off & Restart.*

7. Your device restarts with Lockdown Mode disabled.

8.7 Malware Lessons Learned

☐ Malware is an umbrella term that includes worms, Trojan horses, phishing attacks, malicious scripts, spyware, viruses, etc.

☐ There are from 500,000-40,000,000 malware in the field, with up to 40,000 new signatures appearing daily.

☐ Although Windows-specific malware cannot impact macOS, it can be inadvertently passed along to a Windows device via email or file sharing.

☐ AVComparatives.com is an independent anti-malware testing lab with public reports on the effectiveness of commercial anti-malware products.

☐ Bitdefender is one of the consistently top-performing anti-malware products. It is cross-platform, inexpensive, and easy to install and configure.

☐ To help ensure your software is not compromised, only download software from the Apple App Store or the developer's website.

☐ New as of macOS 13 is Lockdown Mode, which adds extreme cybersecurity measures to iPhone, iPad, and macOS.

8.8 Additional Reading

Souppaya, Murugiah, and Karen Scarfone. "Guide to Malware Incident Prevention and Handling for Desktops and Laptops." <u>NIST Special Publication 800-83, Revision 1</u>. July 2013. <http://nvlpubs.nist.gov/nistpubs/SpecialPublications/NIST.SP.800-83r1.pdf>

9 Firewall

Expecting the world to treat you fairly because you are a good person is a little like expecting the bull not to attack you because you are a vegetarian.

–Dennis Wholey[1]

What You Will Learn in This Chapter

- What a firewall does
- Activate the firewall
- Close unnecessary ports

What You Will Need in This Chapter

- No additional resources required.

9.1   Firewall

Whenever a computer, mobile device, or network device needs to communicate with the outside world–say, to print, receive or send email, or surf the web–it must *open a door* to that world. In the IT universe, this is called *opening a port*.

Ports are numbered from 1 to 65,535, with at least one unique port number assigned to any one communication task. For example, when using your browser to visit Google, you enter `https://www.google.com` in the address field. This can be translated into English as: *Using the language of the secure Internet*

---

[1] *https://en.wikipedia.org/wiki/Dennis_Wholey*

(https) I would like to communicate with a server named www within a domain named google.com.

The problem is that the www server at Google has 65,535 ports to which it may potentially need to listen. Invisible to the user, *:80* is placed at the end of the address request. This translates into: *And please knock on port 80 (reserved for web server communications) so that www can respond to the web page requests sent to it.*

To best secure your computer, it is important to only have those ports open that are necessary to perform your work.

The purpose of a firewall[2] is to block unwanted attempts to get into or communicate with your computer from the network or Internet through your 65,535 ports. It is about as simple as anything can be on a computer, and once activated you never need to know about it again.

To get into your computer or to communicate with it, the following must be true:

- Your computer must be on a network with other computers (such as your local area network at home or office, or the Internet).

- You must have a port open. On macOS, ports are opened by enabling sharing services from the Sharing System Setting and by some applications.

- Lastly, there must be some process or application listening at the port that can respond. You can open port 80 on a web server, but if the web server application (typically Apache) hasn't been launched, no amount of your browser screaming at the server elicits a response.

In macOS, activating the firewall puts guards at the gates to prevent unwanted visitors.

The second step is to close those ports whose associated services you do not need. This is accomplished by disabling unnecessary services in *System Settings > Sharing*.

9.1.1   Assignment: Activate the Firewall

In this assignment, you enable the built-in macOS firewall.

---

[2] *http://en.wikipedia.org/wiki/Firewall*

1. Open *Apple* menu > *System Settings* > *Network* > *Firewall*.

2. Enable *Firewall*.

3. Select *Options…* button to further refine the firewall.

4. The *Firewall Options* window opens:

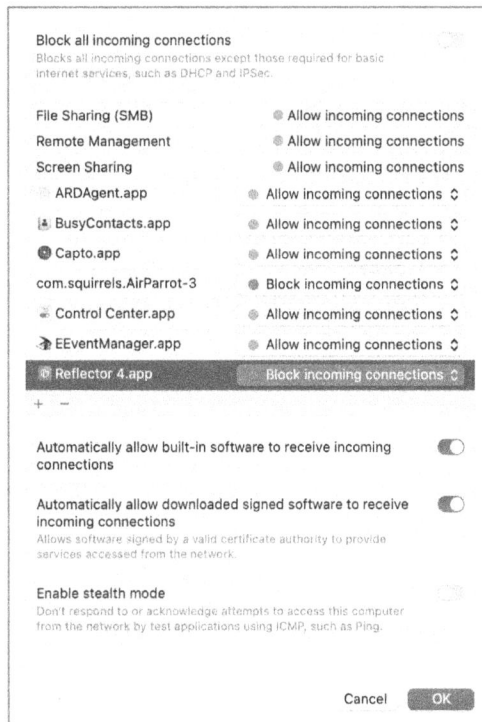

- Enabling *Block all incoming connections* effectively closes all but the most essential ports. You still can reach out to initiate communications, such as to surf on a web browser or initiate a call on Skype, but you must do the initiating or others will not get in. Unless you are working in an unusual environment, this step is not necessary.

- In the main body area, you see all the ports that are open. Above the horizontal line are those ports you opened when enabling items in the Sharing System Preference. Below the line are those ports opened by launching applications that require communication outside of your computer. It is possible that some of these applications have no need for outside communications. In that case, you can tap on the *Allow incoming connections*, then select *Block incoming connections*.

- *Automatically allow built-in software to receive incoming connections* is enabled by default. This allows Apple software that is built into macOS to receive communications.

- *Automatically allow downloaded signed software to receive incoming connections* is enabled by default. Applications that are "signed" have special coding that allows macOS to determine if it has been damaged or modified from the original in any way. Apple has given its seal of approval to the original as being free of any intrusion software.

- *Enable Stealth mode* can usually be left disabled. Enabling this checkbox makes your Mac unresponsive to Ping and other network diagnostic tools. However, enabling this checkbox should have no impact on any aspect of your computer use.

5. Select the *OK* button.

6. Quit System Settings.

Congratulations! Your firewall is now on guard, preventing unwanted penetration of your computer.

9.1.2   Assignment: Close Unnecessary Ports

In this assignment, you examine the currently open ports, then close those that are not necessary.

Open Sharing Preferences

1. Open *Apple* menu > *System Settings* > *General* > *Sharing*.

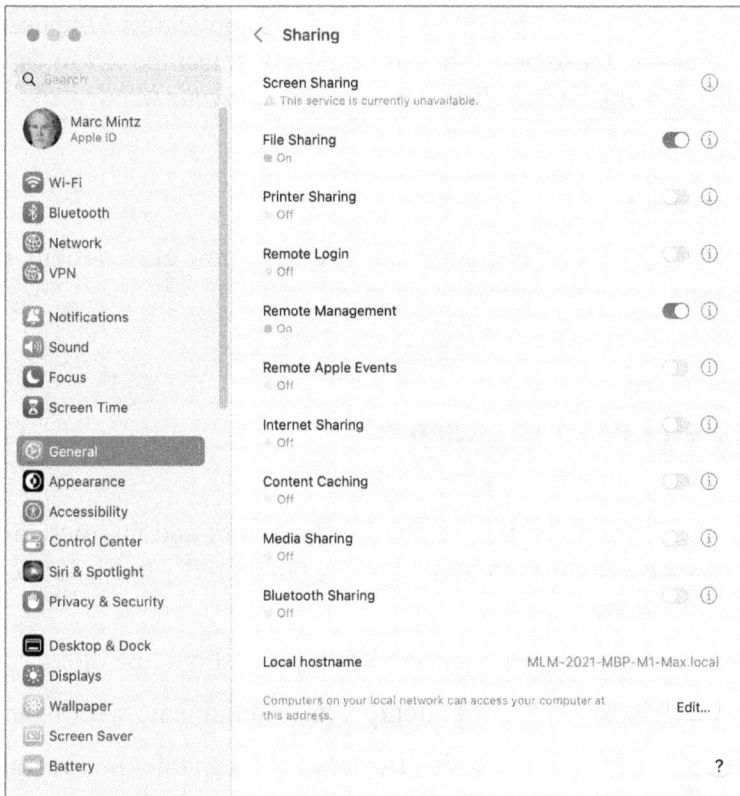

2. If any of your sharing items are enabled, at least one port has been opened to allow communication for that item.

3. If any currently enabled item is not needed, disable its checkbox.

Check Which Ports Are Opened Via User-Launched Applications

4. Select *System Settings > Network > Firewall.*

5. Select the *Options…* button.

6. The Firewall Options window opens, displaying all open ports based on the name of the process or application that has opened them. In this example, above the line I show *Remote Management,* and *Screen Sharing* ports are open.

7. If any of the above the line ports should be closed, return to *System Settings > General > Sharing,* then turn off the shared item.

8. Below the line are ports opened by some of the applications the user has installed. If any of these ports should be closed, select the *Allow incoming connections* pop-up menu > *Block incoming connections*.

9. Tap the *OK* button.

10. Quit System Settings.

If any unnecessary ports had been opened, you have now closed them, preventing unwanted access.

9.2 Firewall Lessons Learned

☐ Other devices and services on the local network and internet communicate with your device through logical ports, numbered 1 to 65,535, each with a unique function.

☐ To best secure a device, only necessary ports should be open.

☐ A firewall blocks unwanted attempts to communicate with your device.

☐ The macOS firewall is inactive by default. To enable the firewall, go to *System Settings* > *Network* > *Firewall*.

☐ Specific operating system and application communication ports can be enabled or disabled from *System Settings* > *General* > *Sharing* and *System Settings* > *Network* > *Firewall* > *Options*.

10 Lost or Stolen Device

It takes considerable knowledge just to realize the extent of your own ignorance.

–Thomas Sowell[1], American economist, social theorist, political philosopher, and currently Senior Fellow at the Hoover Institution, Stanford University[2]

What You Will Learn in This Chapter

- Enable and configure Find My app
- Use Find My from a computer
- Use Find My from an iPhone or iPad

What You Will Need in This Chapter

- No additional resources required.

10.1 Find My Mac App

Millions of computers are stolen each year. If you followed the steps to enable *FileVault 2* with strong login passwords, all a thief gets is the device–not the data.

But it would be nice to be able to get your Mac back.

When your missing/stolen Mac powers on and connects with a Wi-Fi or Ethernet network that connects to the Internet, the device can discover its geographical

---

[1] *https://en.wikipedia.org/wiki/Thomas_Sowell*
[2] *This observation has since been validated in University studies, originally performed by David Dunning and Justin Kruger. It is now referred to as the Dunning-Kruger effect. https://en.wikipedia.org/wiki/Dunning–Kruger_effect*

location. If the device powers on but is not connected to a Wi-Fi or Ethernet network that connects to the Internet, your device will occasionally send out a Bluetooth beacon that can be picked up from nearby Apple devices (Mac, iPhone, and iPad). These devices know their locations. They relay their locations and device names to Apple. When you open the *Find My* app on another of your Apple devices, or log into icloud.com, the location of your missing device is displayed on a map.

For *Find My* to function, the following must happen:

- An iCloud account is activated.
- *Find My Mac* is enabled for the computer.
- The computer is turned on, even if sleeping.
- Another Apple device is active and within Bluetooth range.

10.1.1 Assignment: Activate and Configure Find My Mac

In this assignment, you activate and configure Find My Mac

Enable Find My Mac

1. Select the *Apple* menu > *System Settings* > *Apple ID* > *Find My Mac*.

2. Turn on *Find My Mac*.

3. Turn on *Find My Network*.

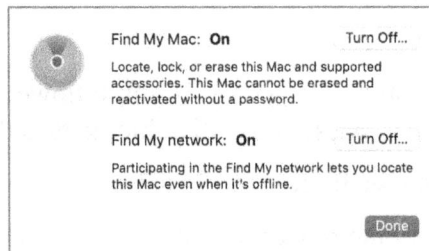

4. Tap *Done*.

Enable Location Services

Allow the device to broadcast its location.

5. Select *System Settings* > *Privacy & Security* > *Location Services*.

6. Enable *Location Services*.

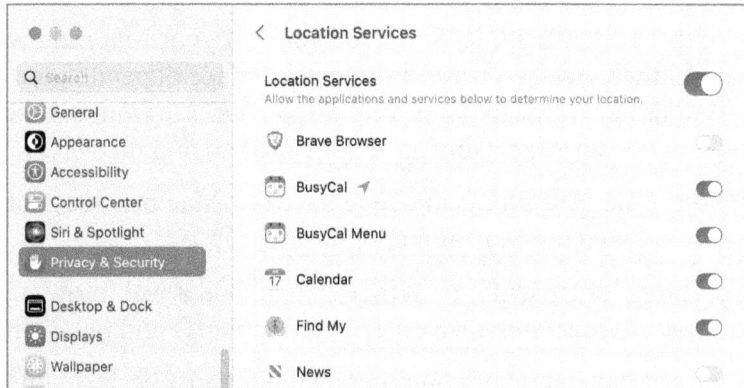

7. Enable *Find My*. Configure the other options to taste.

8. Scroll down to *System Services,* then tap *Details*. My settings are shown below:

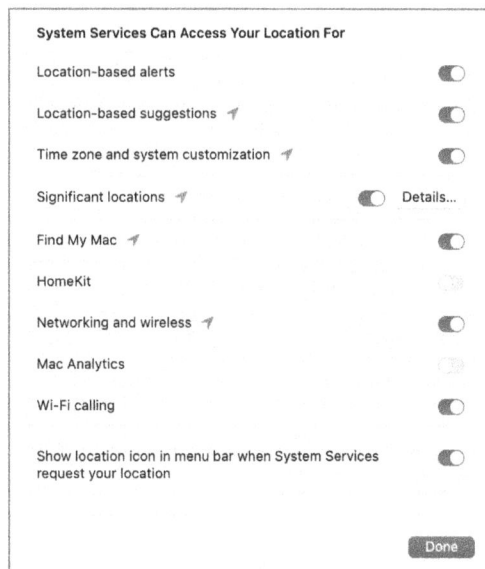

9. Tap the *Done* button.

10.1.2 Assignment: Use Find My from a Computer

For purposes of this assignment, let's assume you lost your Mac, and we use *Find My* to locate it. As there are two ways this can take place depending on the device, we have two assignments: Locate your Mac using *Find My* from another computer, and locate your Mac using an iPhone or iPad and the *Find iPhone* app.

1. Power on your Mac. Assuming someone has your Mac but does not know a login password, leave the Mac at the login screen.

2. On another computer (macOS, or Windows), launch a web browser, visit iCloud at *https://icloud.com,* then enter your Apple ID and password.

    *   If you do not have another computer available, just perform this exercise on your own computer logged in as Guest.

3. If you have previously enabled Apple Two-Factor Authentication, enter the 6-digit code received on any of your authorized devices.

4. The iCloud desktop appears. Select the *Find iPhone* button.

5. The *Find My iPhone* map appears.

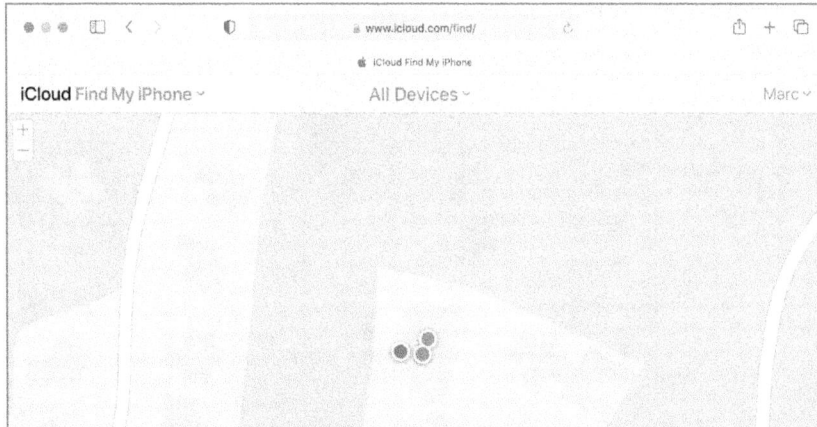

6. At the top, select the *All Devices* menu. All your registered devices appear. If the device is powered on, it has a green light next to it.

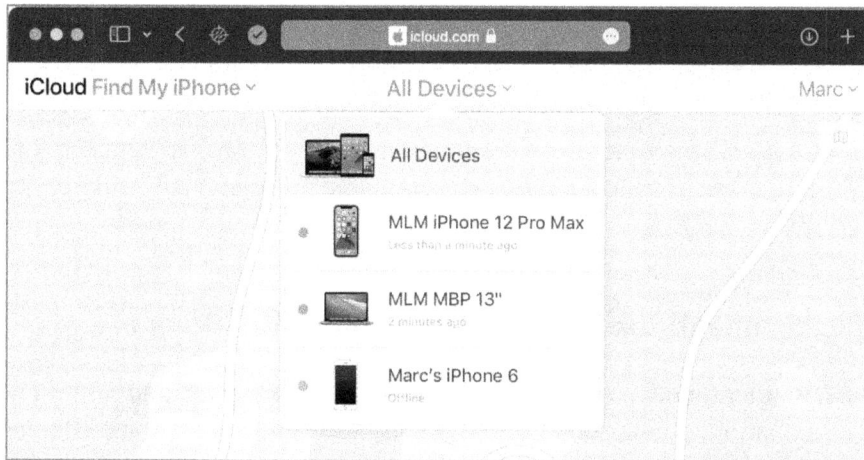

7. From *All Devices*, select the device to be located. Zooming into the map provides a detailed location.

8. Once located on the map, a pop-up window with the name of your device appears. You have access to three options.

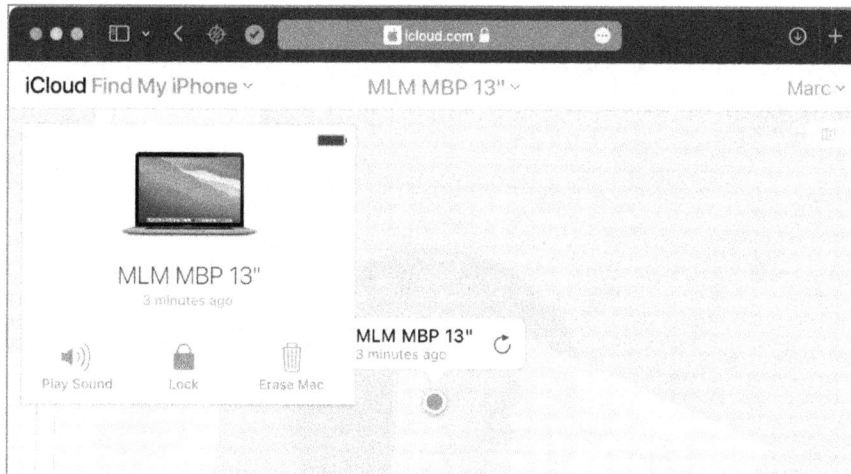

- *Play Sound*. Plays a "sonar" sound and displays an alert window on the device.

- *Lock*. Locks down your system, preventing any use.

I do not recommend either the Play Sound or Lock options, as it notifies the thief you are tracking them. They may turn off the computer, blocking any future tracking, or worse, destroy the "tracking device."

- *Erase Mac*. If it is not possible to get prompt police intervention, you may want to erase your Mac. After all, you have a full current backup, don't you?

10.1.3 Assignment: Use Find My Mac from an iOS Device

For this assignment, you assume the thief took your only Mac, and you do not have access to another computer. However, you do have access to an iPhone or iPad.

- Prerequisite: iPhone or iPad.

1. If you do not already have the *Find My* app installed on your device, visit the App Store and download it.

2. Open the *Find My iPhone* app.

3. Enter your *Apple ID* email address and *password*, then select the *Go* key.

4. The *Find My iPhone* screen opens. Select the target device to locate.

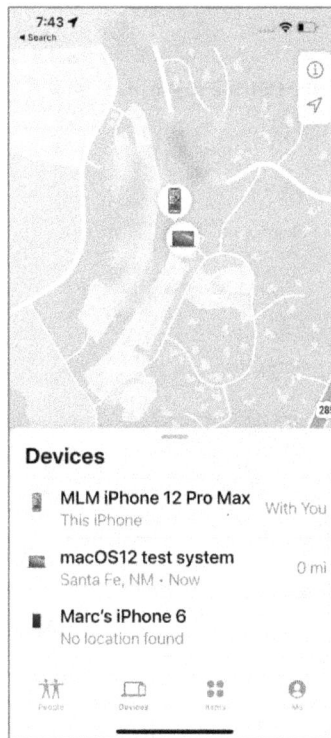

5. The *Find My iPhone* map opens. Tap the name of the target device. The last known location for the target device displays–including the street address!

6. Slide up the *Info Screen*. From here you may have the device play a sonar sound (I use this at least once a week because my yellow Labrador is obsessed with hiding my iPhone), lock it to prevent any access, or erase the drive.

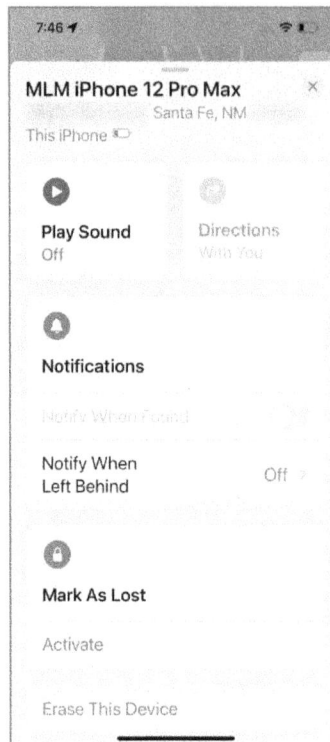

7.   When done, exit the *Find My iPhone* app.

Hooray! You've found your lost/stolen device.

10.2 Bitdefender Anti-Theft

If you completed the previous Malware chapter, you already have Bitdefender
Anti-Virus installed. A benefit of using Bitdefender is that the subscription
includes an *Anti-Theft* device locator. Although Bitdefender *Anti-Theft* works
almost as well as Apple's *Find My*, it only does so on Android, Chrome, and
Windows devices. Not on macOS.

10.2.1 [Android, Chrome, and Windows] Assignment: Find a Device from a Computer

10.3 Lost or Stolen Device Lessons Learned

☐ *Find My Mac* is built into macOS. Once enabled, it is possible to locate your device if powered on.

☐ Enable *Find My Mac* from *System Settings > Apple ID >* enable *Find My Mac* checkbox > enable *Find My Mac* and *Find My network*.

☐ Bitdefender *Anti-Theft* does not work with macOS.

11 Local Network and Bluetooth

I am concerned for the security of our great Nation; not so much because of any threat from without, but because of the insidious forces working from within.

–General Douglas MacArthur[1]

What You Will Learn in This Chapter

- Block Ethernet broadcasting
- Prevent Ethernet insertion
- Understand Wi-Fi encryption protocols
- Configure router security
- Secure Bluetooth devices

What You Will Need in This Chapter

- No additional resources required.

11.1 Ethernet Broadcasting

It is common wisdom that Ethernet is more secure than Wi-Fi. But as with most things we believe, this is not accurate.

There are two security issues with Ethernet: Broadcasting and Insertion. At the most fundamental level, what happens when data travels through Ethernet is that electrons travel along a metal cable. There are two unintended consequences that

---

[1] *https://en.wikipedia.org/wiki/Douglas_MacArthur*

occur whenever electrons go for a ride: heat generation, and creation of an electromagnetic field. For our purposes, heat is not an issue. But the electromagnetic field is.

Sending data through copper wire effectively turns that wire into a very large antenna that broadcasts your data through radio waves. With the right receiver and translation software, you can easily capture every bit of data being sent and received along that cable.

This vulnerability is not something about which the average person or business would or should be concerned. On the other hand, if you or your business requires the utmost in security, it is mandatory to add encryption to your Ethernet network. This can be done by using a *Virtual Private Network* (*VPN*) service or installing a RADIUS server on your network. VPN is discussed in the Internet chapter. RADIUS use is beyond the scope of this book.

11.2 Ethernet Insertion

You would notice if someone came into your home, plugged a computer into your network, and sat there watching data go by. But in a typical business, nobody would notice.

Ethernet and Wi-Fi networks can be protected from unwanted insertions by implementing the 802.1x protocol[2] (often referred to as RADIUS). This protocol works with both Ethernet and Wi-Fi, mandating that anyone attempting to join the network authenticate with their own personal name and password. This is unlike the typical Wi-Fi authentication that uses the same password for everyone.

To implement 802.1x you need to have either: a macOS, Windows, or Linux Server running within your network; one of the many other 802.1x appliances that are sold; or even a cloud service that provides RADIUS service[3]. Details on how to configure 802.1x are beyond the scope of this book.

---

[2] *https://en.wikipedia.org/wiki/IEEE_802.1X*
[3] *https://jumpcloud.com/blog/mac-radius-authentication*

11.3 Wi-Fi Encryption Protocols

Right out of the box, many Wi-Fi base stations are insecure. Anyone who can pick up the signal can connect. This allows them to not only use your bandwidth to access the Internet, but also to see all the other data such as usernames and passwords that travel on that network. To start securing your Wi-Fi, add strong password protection with encryption.

Although cellular networks do use encryption, the protocol in use has been broken for many years, making it easy for a novice hacker to see all data passing though. In addition, it is common practice for police and other government agencies to set up their own cellular towers[4] (Stingray) with the purpose of harvesting data.

To prevent your data from being seen while on a cellular network or an insecure Wi-Fi network, it is necessary to use VPN (Virtual Private Network) encryption (more on that later). If the Wi-Fi network is properly encrypted, you should have little concern about the security and privacy of your data.

Below is a brief on each of the Wi-Fi encryption protocols.

- **WEP**[5] (Wired Equivalency Protocol) was the first encryption protocol for Wi-Fi. Introduced in 1999, it was quickly broken, and by 2003 was replaced by WPA and WPA2 (Wi-Fi Protected Access). Any Wi-Fi base station manufactured in the past 5 years offers WPA and WPA2 in addition to WEP.

There is only one reason to ever use WEP: you simply have no other option. Kids driving by your home can break into your WEP network before leaving the block.

- **WPA**[6] (Wi-Fi Protected Access) superseded WEP in 2003. Although it is a great advancement, it too has been broken. As with WEP, the only reason to use WPA is that you have no other option.

- **WPA2**[7] superseded WPA in 2004. Although technically difficult, WPA2 can also be hacked.

---

[4] *https://en.wikipedia.org/wiki/Stingray_phone_tracker*
[5] *http://en.wikipedia.org/wiki/Wired_Equivalent_Privacy*
[6] *http://en.wikipedia.org/wiki/Wi-Fi_Protected_Access*
[7] *http://en.wikipedia.org/wiki/Wi-Fi_Protected_Access*

- **WPA3** Superseded WPA2 with certification beginning June 2018. All wi-fi devices certified for WPA after June 30, 2020, must include support for WPA3.

Two encryption algorithms are available to you: *TKIP* and *AES* (technically known as CCMP, but virtually all vendors refer to it as AES). TKIP has been compromised and is no longer recommended. If your Wi-Fi device allows the option of AES, use only that. If it only allows for TKIP, trash the unit, then purchase a more modern device.

11.4 Network Device Overview

The connection point between your Internet Service Provider (ISP) and your Local Area Network (LAN) is a router. A router is a device designed to connect two different types of networks and provide resources for them to interact.

Common brands of routers include Cisco, Ubiquiti, Linksys, Netgear, D-Link, and many unbranded devices that Internet Service Providers lease to customers.

Some newer routers, especially those provided by ISPs, are all-in-one units containing several if not all the components below:

- **Modem**[8]. This hardware decodes and modulates the signal from your Internet provider to your cable or telephone jack. A modem is likely to be a separate component if more than one device exists for your Internet connection.

- **Router**[9]. This component runs a specialized program that allows hundreds of different devices to interact on a network, usually sharing a single IP address to the Internet. Routers use *Network Address Translation* (NAT) to convert and direct Internet traffic from websites to your computer and from your computer to other computers and peripherals on the *Local Area Network* (LAN).

---

[8] *https://en.wikipedia.org/wiki/Modem*
[9] *https://en.wikipedia.org/wiki/Router_(computing)*

- **Firewall**[10]. A firewall may be either software or hardware. Most modern routers now include a firewall, as do most operating systems. The firewall inspects data traffic between the internet and internally connected devices.

- **Intrusion Detection System** and **Intrusion Prevention System**[11]. These features are built into high-end routers. They perform deep packet inspection to further protect the network from outside intruders.

- **Network Switch**[12]. This hardware component allows multiple devices to be connected simultaneously and interact with the router using Ethernet cables. A switch is the modern version of an ethernet hub. Hubs present a critical security vulnerability and should never be used.

- **Access Point**[13]. An access point hardware component allows tens or hundreds of wireless (Wi-Fi) devices to connect to it.

Every router has at least some basic security controls built in. These controls include the ability to filter out what the router thinks are attempts to hack into your network, and the ability to forward specific types of data packets to a specific computer in your LAN, or to point specific types of data packets to a specific computer on the Internet.

Malware, hackers, criminals, and even some government agencies, sometimes attempt to alter security control configurations so that the malware or perpetrators have an easier time harvesting your data. Because of this, it is wise to routinely inspect the condition of your router. How often is *routine?* Within larger or security-conscious organizations with high-value data, it is common to have a network administrator dedicated to maintaining watch over the status of network equipment. For a small business or household, once a month is not too often.

---

[10] *https://en.wikipedia.org/wiki/Firewall_(computing)*
[11] *https://en.wikipedia.org/wiki/Deep_packet_inspection*
[12] *https://en.wikipedia.org/wiki/Network_switch*
[13] *https://en.wikipedia.org/wiki/Wireless_access_point*

11.4.1 Assignment: Determine Your Wi-Fi Encryption Protocol

It is vital to know if the Wi-Fi network your device connects to is secure with WPA2 or WPA3. If it is not, almost *everything* your device does on that network may be viewed-including usernames and passwords.

In this assignment, you determine the Wi-Fi encryption protocol that is in use.

● Prerequisite: An available Wi-Fi network for which you know the password.

1. Tap the *Wi-Fi* icon in the menu bar.

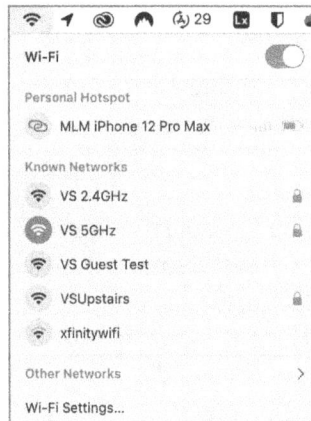

2. If there is a *lock* icon to the right of the target network name, the Wi-Fi is encrypted. However, from this view there is no way to know if it is WEP, WPA, WPA2, or WPA3.

3. If this is the network you are currently connected to, hold down the *Option* key then tap the Wi-Fi menu icon to display a detailed view. Included in this view is the security protocol:

4. If this is a network you are not currently connected to, select *Apple* menu > *System Settings* > *Network* > *Wi-Fi*. All available networks are displayed:

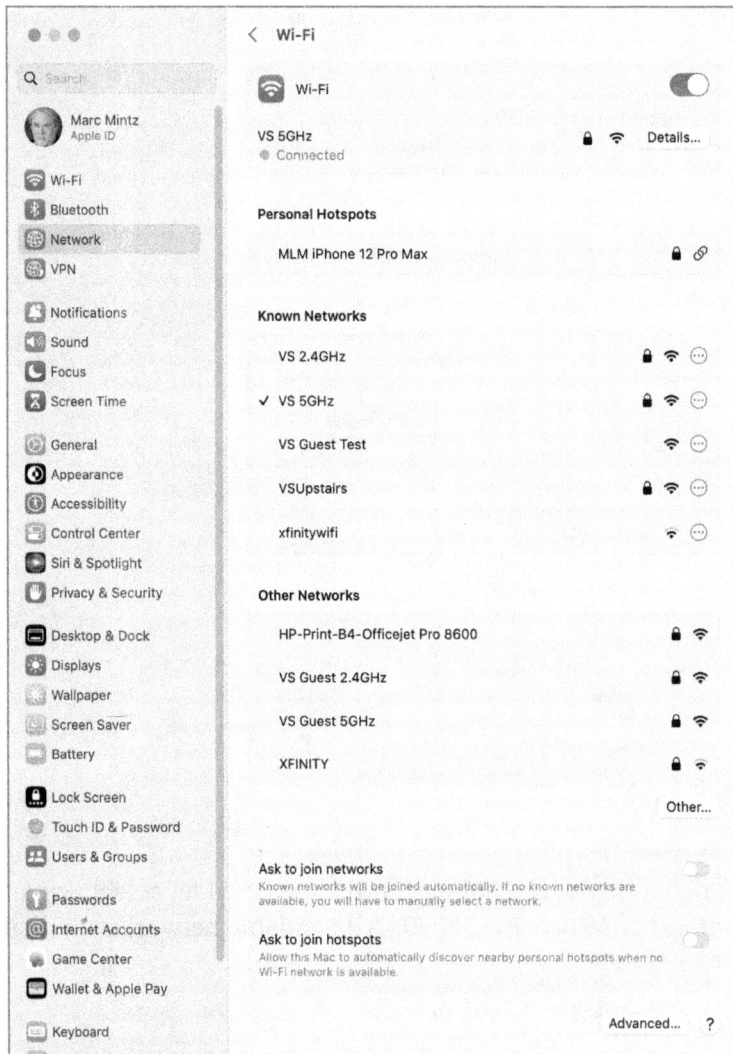

5. Select the *Advanced* button from the bottom right corner to display the Wi-Fi networks you have connected with, along with their security protocols:

Require administrator authorization to

Change networks

Turn Wi-Fi on or off

Show legacy networks and options

Wi-Fi MAC address   b0:de:28:41:d5:03

Known Networks

Network Name	Security Type	
🔒 AdamsCrow-Guest	WPA3 Personal	
BROADCOM_GUEST_...	None	
Caesars_Resorts	None	
🔒 Days_Inn_ABQ_North...	WPA3 Personal	
FairfieldInn_GUEST	None	
McCarran WiFi	None	
🔒 NinePrinces	WPA2 Personal	
🔒 SFPLSS	WPA2 Personal	
🔒 SOUTHWEST GUEST	WPA3 Personal	
🔒 VS 2.4GHz	WPA3 Personal	
🔒 VS 5GHz	WPA3 Personal	
VS Guest Test	None	
🔒 VSUpstairs	WPA3 Personal	
xfinitywifi	None	

Done

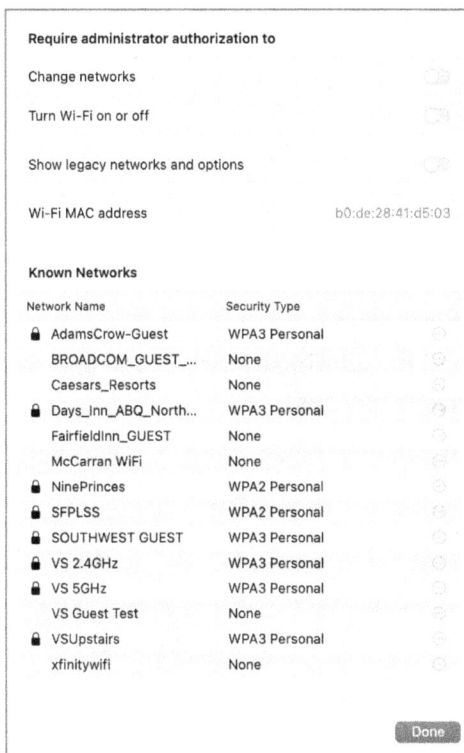

6. If connecting to a Wi-Fi network for the first time, the authentication window will display the encryption protocol. If no authentication window appears and you do connect with the network, there is no encryption.

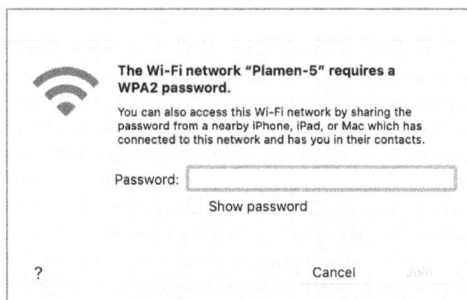

The Wi-Fi network "Plamen-5" requires a WPA2 password.

You can also access this Wi-Fi network by sharing the password from a nearby iPhone, iPad, or Mac which has connected to this network and has you in their contacts.

Password: [_____]

Show password

?   Cancel   Join

If the protocol is WPA2 or WPA3, life is all rainbows and unicorns. If it is anything else, *everything* you do on that network can be clearly visible to others. I strongly recommend not using this network unless you have installed *VPN* software to encrypt your Internet traffic (more on this later).

11.4.2 Assignment: Configure WPA2 or WPA3 on Your Router

This assignment is generic, as each router has its own unique interface to configure Wi-Fi encryption.

In this assignment, you configure your Wi-Fi router to use the secure WPA2 or WPA3 protocol.

- Note: If this assignment is performed in a class, the instructor demonstrates while the students observe.

- Prerequisite: You need the administrator's name and password of the router to view and edit the current settings.

- Prerequisite: You need the user manual for your router.

Find the IP address of your Wi-Fi router

1. While holding down the *Option* key, tap on the *Wi-Fi* menu icon. A detailed list of your Wi-Fi stats displays, including your router address.

2. Locate the *Router* address. This is your Wi-Fi base station or router IP address. Write this down to remember it.

3. Close the pane.

Access the router

4. Open a web browser to the IP address of the router.

5. At the *Authentication* window, enter the administrator's name and password. This is the administrator of the router, not of your computer.

6. The router control panel appears.

● Note: Below is a screenshot of an ASUS GT-AXE11000 router. Each router has its own unique interface.

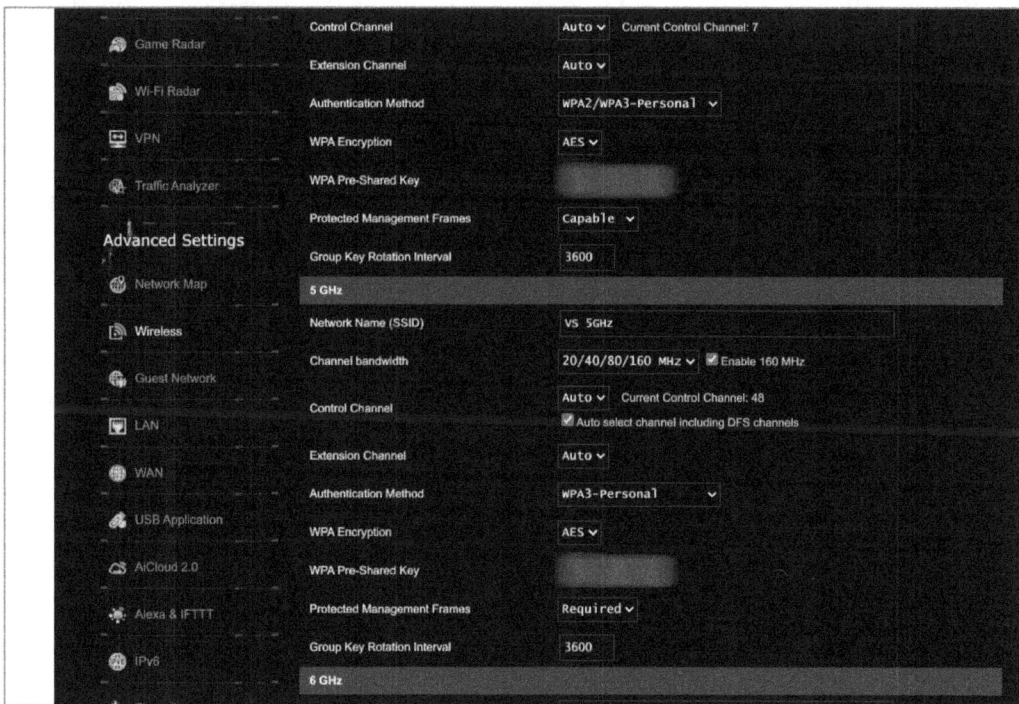

7. Follow the instructions in the router manual to configure Wi-Fi encryption to WPA2, or if available, WPA3.

   ● **Warning**: If your router has the option of using either *AES* or *TKIP*, select *AES*. If you only have an option for TKIP, immediately replace this router. The TKIP encryption scheme has been broken and is easily hacked.

8. If any changes were made, tap the *Apply* button to save the changes.

9. Close the browser window to exit out of your router.

Congratulations! All traffic on your Wi-Fi now is securely encrypted.

11.5 Use MAC Address To Limit Wi-Fi Access

Every device that can connect to a TCP network has a unique *MAC Address*[14] (Media Access Control), sometimes referred to as the *Hardware Address*. This address specifies the manufacturer of the device, and a device-specific number. Do not go to sleep on me yet! This MAC address can be used with most Wi-Fi base stations to limit what devices can connect to your network.

Although every Wi-Fi base station has a unique interface to filter by MAC address, they all operate on the same principle: either allow anyone with the proper password to gain access to the network *or* allow anyone with the proper password *and* proper MAC address access to the network. In this way, you can easily lock down your Wi-Fi to only approved devices. So even if employees know the password, they are unable to connect their personal device to the Wi-Fi unless the MAC address for those devices is on the list.

- Note: Consider MAC Address Filtering as something to keep the kids out of the network, not skilled hackers. It is possible to see the MAC addresses of some network-connected devices, then to temporarily fake having one of those MAC addresses, which can help a hacker get onto the network.

11.5.1 Assignment: Restrict Access by MAC Address

In this assignment, you configure a router to allow only desired devices to connect. This assignment is generic, as each router has its own unique interface.

- Note: If this assignment is performed in a class, the instructor demonstrates while the students observe.

- Prerequisite: You need the administrator's name and password of the router to view and edit the current settings.

---

[14] *http://en.wikipedia.org/wiki/MAC_address*

- Prerequisite: You need the user manual for your router.

Find and record the IP address of your permitted wireless devices

1. Make a list of devices to be permitted access to your Wi-Fi network. Include a 1 or 2-word description, and the MAC address of the device. As an example:

Device Name	User	Description	MAC
Roku TV	Staff	58" Conference room	XX:XX:XX:XX:XX
MacBook	Marc	2020 13" M1	XX:XX:XX:XX:XX

- The MAC address of an Android device may be found from *Settings > Connections >* tap your *current network name >* tap *Gear* icon next to the current network name > *Advanced >* scroll to the bottom of screen for *MAC address.*

  - Note: The *MAC address type* must be set to *Phone MAC,* not *Randomized MAC,* or it constantly changes.

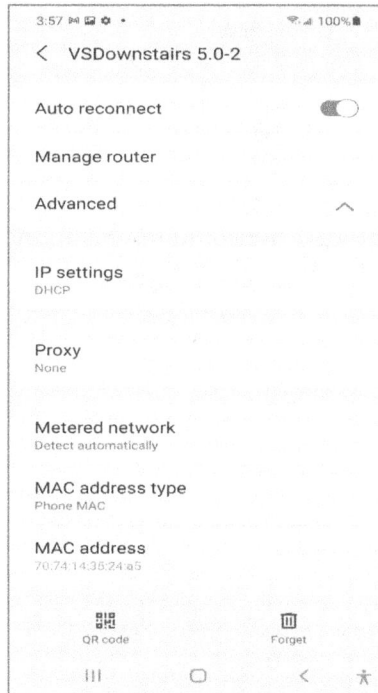

- The MAC address of a Chrome OS devices may be found from *Settings > Network > current network name > current network name > expand Network area > MAC address* field.

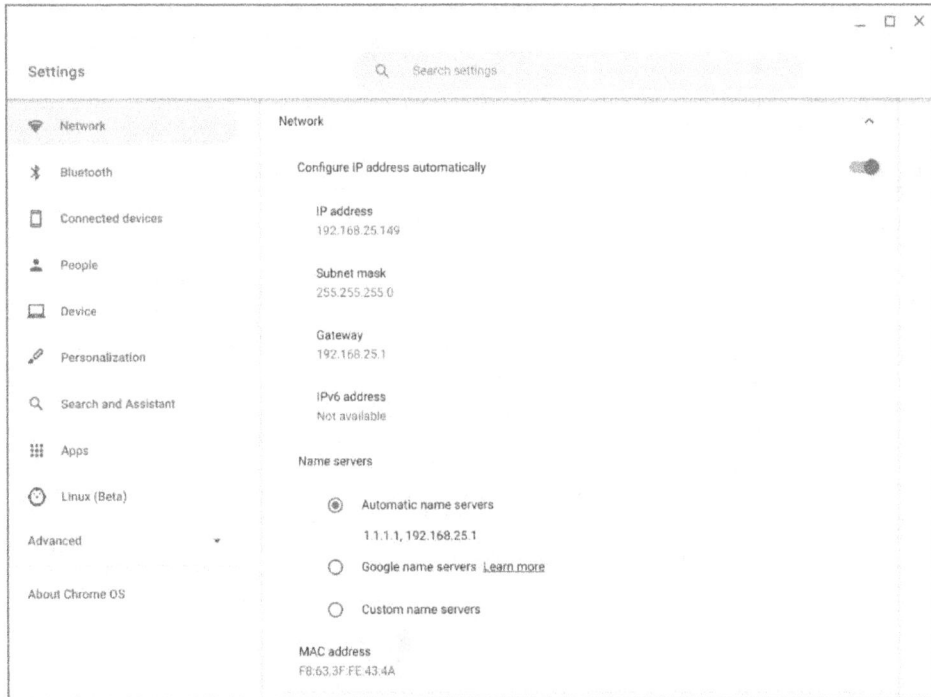

- The MAC address of an iPhone and iPad is found in the *Settings > General > About > Wi-Fi Address* field.

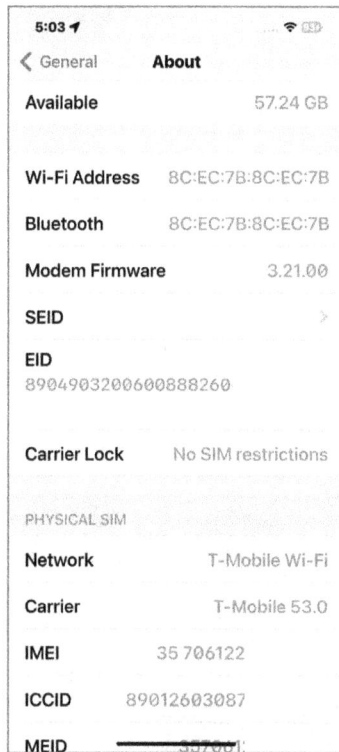

- The MAC address of a macOS 13 device may be found from the *Apple* menu > *System Settings* > *Network* > connected network > *Details* button > *Hardware* > *MAC address*:

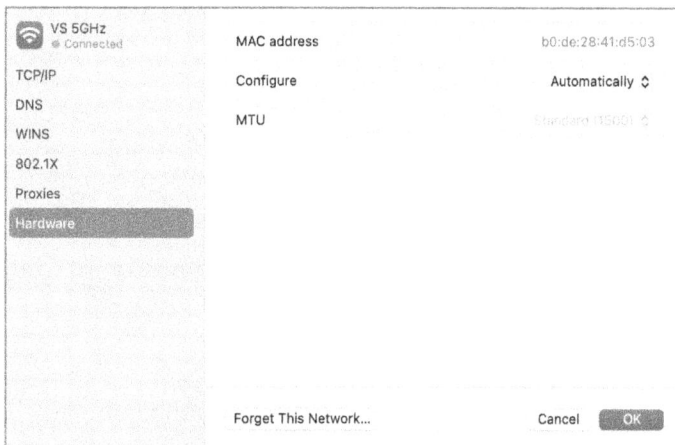

- The MAC address of a Windows 11 device is found with the *ipconfig* command in the command prompt.

a. Tap the *Start* button in the Taskbar.

b. Type cmd, then tap the *Command Prompt app*.

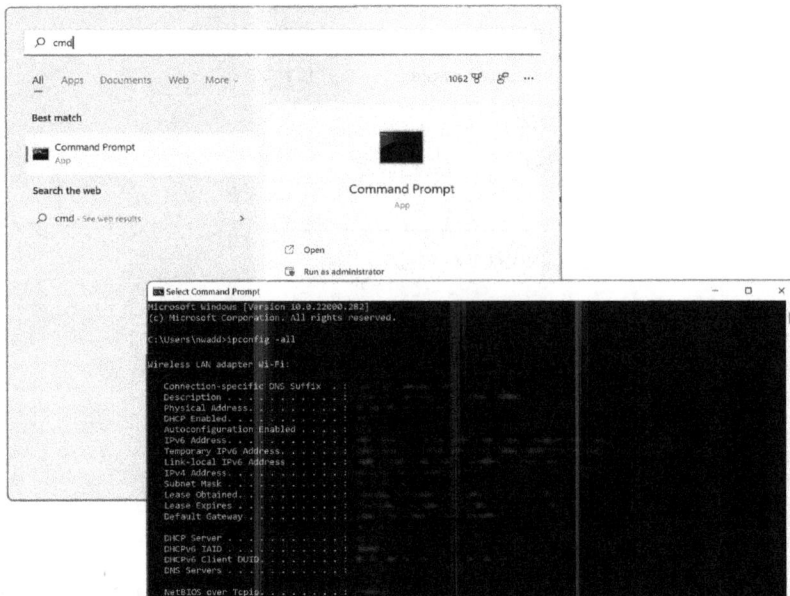

c. Type ipconfig -all into the command line, then tap *Enter*. A listing of all network addresses for the device appears. The MAC address shows as the *Physical Address*.

d. Close the Command Prompt.

Configure the router

2. With your list of desired devices in hand, launch a browser and enter the IP address of your Wi-Fi router.

3. At the authentication window, enter the name and password of the administrator for the router.

4. The router control panel appears.

- Keep in mind that all routers–even from the same company–have slightly different interfaces.

5. Follow the user manual for your router to enter the page for *MAC Filter,* then enter the MAC address and device description for each device to be allowed Wi-Fi access to your network.

6. *Save* changes.

7. Close the browser window to exit out of your wireless router.

Congratulations! You have secured your wireless network so that only authorized devices have access.

11.6 Router Penetration

The connection point between your Internet provider cable, DSL, fiber, radio, etc. and your Local Area Network (LAN) is a router. A router is a device designed to connect two different types of networks.

 Common areas of router penetration include:

- **Port forwarding**[15]: Port forwarding allows a service–such as an internal web server–to be accessible from the internet. However, if ports are being forwarded without purpose, the firewall is being bypassed and your internal computers may be visible from the internet.

- **DMZ**[16]: Related to Port Forwarding is the DMZ, or De-Militarized Zone. DMZ is typically used to route *all* external traffic for a specific IP address to a specific computer, regardless of service request. Unless there is a unique need, it should remain disabled or else it may be used for nefarious purposes.

- **RAM-Resident Malware:** Some router malware make their home in the RAM of the router. In this way, they can take control of your data traffic without showing in the interface.

- **Firmware**[17]: It is vital to keep the router firmware up to date. Just as with any software, router firmware always has vulnerabilities. Over time, criminals

[15] *https://en.wikipedia.org/wiki/Port_forwarding*
[16] *https://en.wikipedia.org/wiki/DMZ_(computing)*
[17] *https://en.wikipedia.org/wiki/Firmware*

(and some government organizations) discover how to use these vulnerabilities to their benefit. Keep the firmware updated to stay a step ahead of this problem.

11.6.1 Assignment: Verify Router Security Configuration

In the example below, I use an ASUS GT-AXE11000. Although all routers have a different interface, most share the same functions.

In this assignment, you verify the security configuration of a router.

- Note: If this assignment is performed in a class, the instructor demonstrates while the students observe.

- Prerequisite: You need the administrator's name and password for your router.

- Prerequisite: You need the user manual for your router.

Remove RAM-resident malware

Some malware make their home in the router RAM. Also, over time the router RAM may accumulate corruption. The fix for both issues is the same: power cycling.

1. Verify that all users have disconnected from the network and Internet and have closed any connections to other devices on the network.

2. Power off the router. If your router does not have an on/off switch, pull the power cord from the back of the router.

3. Remove the router batteries (if any).

4. Wait a minute.

5. Insert the router batteries (if any).

6. Power on the router. It may take up to 5 minutes for it to be fully operational.

7. Open a browser and enter the IP address of your router.

8. At the prompt, enter the administrator's name and password.

Verify router firmware is up to date

9. Follow your router user manual to check if your firmware version is current. If not, follow your router user manual to update the router firmware. The typical steps are to:

   a. From a browser logged into the router control panel, download the current firmware.

   b. From the router control panel, upload the firmware to the router.

   c. From the router control panel, install the new firmware.

   d. Restart the router.

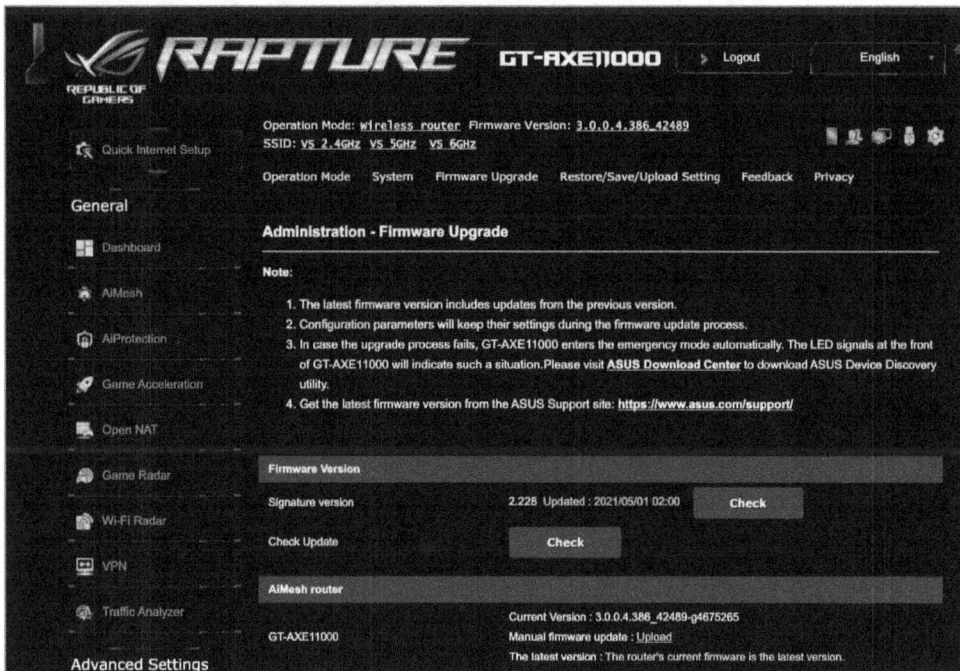

Verify no unnecessary port forwarding

10. Follow your router manual to display the *Port Forwarding* page.

11. Verify *Port Forwarding is* disabled. If it is enabled, verify the need for the activity.

Verify DMZ configuration

DMZ is like *Port Forwarding*. When *DMZ* is enabled, all inbound packets are routed to the specified device. This allows a single device on your network to be accessible from the Internet. Such access presents a high level of vulnerability for that device. Unless there is a demonstrated business need for this function, and adequate steps have been taken to prevent unwanted penetration, turn off *DMZ*.

12. Follow your router manual to display the *DMZ* page.

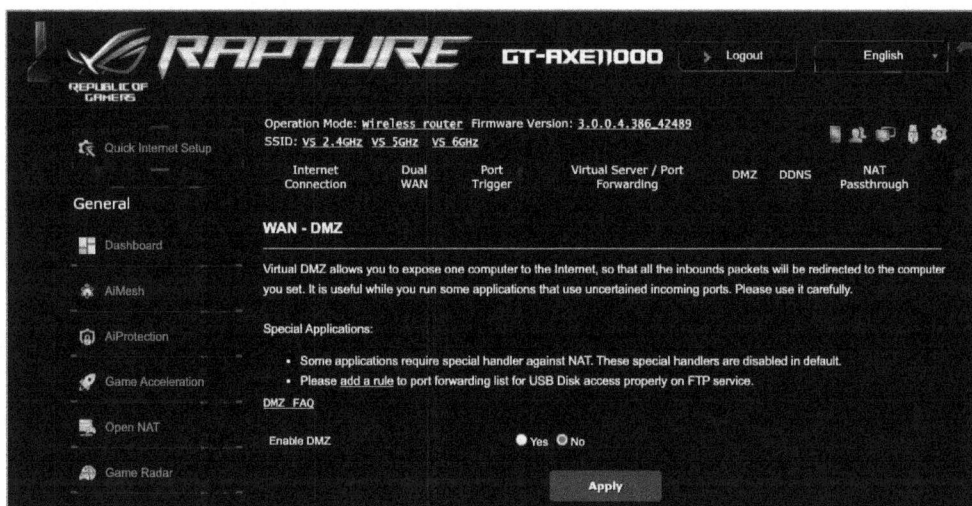

13. Verify *DMZ* is disabled. If it is enabled, verify the need for the activity and disable DMZ if there is no need. In the very rare environment where enabling DMZ is required, understand that all inbound traffic is forwarded to the target device, and that device loses all protection of the router's firewall. The device is fully exposed to the internet and must include tools to protect itself from attack.

14. Exit the browser.

Congratulations, your router is in great shape. Remember to perform this same checkup at least monthly.

11.7 Bluetooth

Bluetooth[18] is a wireless standard like Wi-Fi but intended for short range (typically under 30 feet), and for use with peripheral devices with low bandwidth needs such as mice and keyboards. It uses the 2.4 GHz spectrum, which is shared with older Wi-Fi, microwave ovens, garage door openers, wireless headphones, and wireless landline phones.

---

[18] *https://en.wikipedia.org/wiki/Bluetooth*

All modern computers, mobile devices, and Bluetooth devices require authorization prior to linking (also called pairing) one device to the other. This is usually in the form of entering a unique code on one or both devices.

As Bluetooth devices are typically designed to provide some control over the computer – for example, by using a mouse or keyboard – there is significant vulnerability in allowing an unknown Bluetooth device access to your system. Preventing unapproved access is key to maintaining Bluetooth device security.

11.7.1 Assignment: View Paired Bluetooth Devices

In this assignment, you verify the devices connected to your computer via Bluetooth.

1. Open *System Settings > Bluetooth.*

2. All paired devices display:

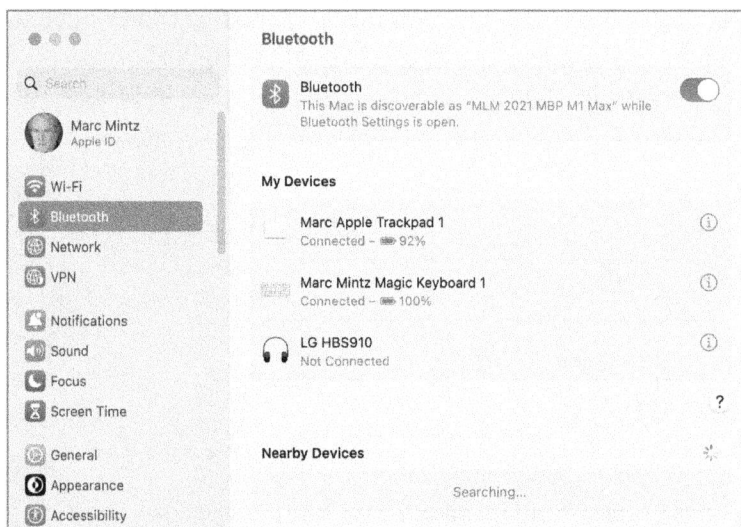

11.7.2 [Optional] Assignment: Add a Bluetooth Device

In this assignment you connect a new Bluetooth device.

•   Prerequisite: A Bluetooth device to connect to your computer.

- **Warning**: Never authorize a Bluetooth connection unless you are trying to pair it for the first time. If an authorization notice appears and you are not attempting to pair a device, it may be a bad actor attempting to connect to your computer.

1. To prevent unauthorized access to your Bluetooth device or computer, adding a Bluetooth device should be performed in an environment away from others– preferably at least 60 feet (although Bluetooth can be manipulated to transmit over one kilometer.

2. To prevent unauthorized access to your computer, do not approve a Bluetooth pairing request unless you are in the process of adding a device.

3. Power on the target Bluetooth device.

4. With computer powered on, open *System Settings > Bluetooth*. Your screen then displays any connected and available Bluetooth devices.

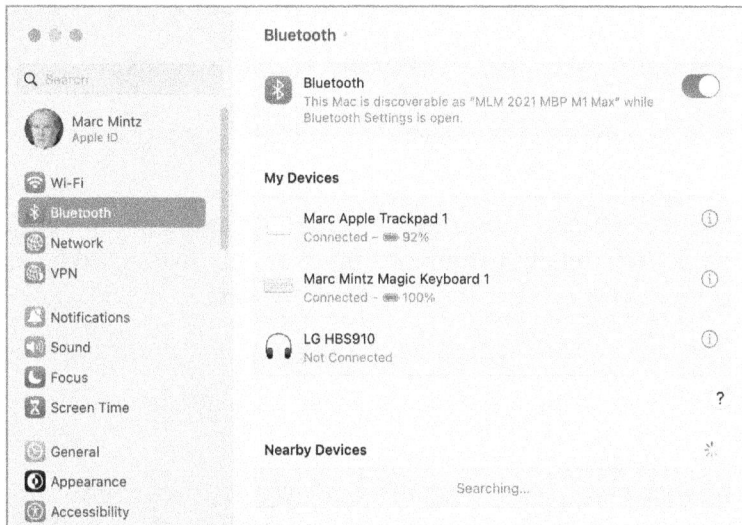

5. To the right of the target Bluetooth device, tap the *Connect* button.

6. Depending on the Bluetooth device, you may be prompted to enter a code. In this case, *0000* is the default code, but you have the option to customize it.

11.7.3 [Optional] Assignment: Remove a Bluetooth Device

Should you no longer need to use a Bluetooth device, it has been lost, or you find an unauthorized device on your Bluetooth list, remove it from your paired device list.

In this assignment, you remove a Bluetooth device from your computer.

1. On your computer, open *System Settings > Bluetooth.*

2. To the far right of the name of the connected device you want to remove, tap the X icon.

3. At the verification prompt, tap *Remove.*

4. The Bluetooth device now is unpaired from the computer, and if the device is within range and powered on, displays a *Connect* button in case you wish to reconnect the device.

11.7.4 Assignment: Security Protocols for Bluetooth

Listed below are the NSA-recommended guidelines[19] for working with Bluetooth devices.

- Choose a device that provides only the features needed to accomplish its purpose, without including extraneous functionality that could be exploited by an attacker who wants access to your device or data.

- Pair the device in a secure area, away from possible eavesdroppers.

- If possible, disable profiles/services and features that are not being actively used.

- Do not authorize connections from new devices or enter/approve a passkey unless you are trying to pair devices for the first time. [You could link to the enemy.]

---

[19] *https://apps.nsa.gov/iaarchive/customcf/openAttachment.cfm?FilePath= /iad/library/ia-guidance/tech-briefs/assets/public/upload/Bluetooth-for- Unclassified-Use-Guidelines-for- Users.pdf&WpKes=aF6woL7fQp3dJirHamRT5DfsLHqKp4xf9Q9ATR*

- Choose a device that clearly communicates its status through audio, text, graphics, and/or LEDs.

- Always maintain physical control over your device. If you lose a device, remove it as soon as possible from the paired device list.

- If applicable to the device, apply patches regularly, use device firewalls, and keep anti-virus software up to date.

- If working in an enterprise setting, comply with all Bluetooth policies and guidance for that enterprise.

11.8 Local Network and Bluetooth Lessons Learned

☐ Ethernet broadcasting describes how data passing through ethernet cabling emits an electromagnetic field that can be read from a distance. This is a reason why all traffic in and out of your devices should be encrypted.

☐ Ethernet insertion is when an unauthorized device is connected to your ethernet network. This is a reason why only necessary ethernet ports should be active.

☐ RADIUS (the 802.1x protocol) is a strategy to encrypt all ethernet and Wi-Fi traffic on your network and devices.

☐ WEP is a legacy (old) Wi-Fi encryption protocol that should not be used.

☐ WPA is a modern Wi-Fi encryption protocol, currently at WPA3. WPA2 and WPA3 are the only versions to be used because the original WPA is no longer secure.

☐ A modem is a hardware device the encodes and decodes and modulates the signal between your internet provider and your network.

☐ A router allows multiple devices to interact on a local network and internet.

☐ A firewall is software or hardware that inspects data traffic between the internet and a device.

☐ Intrusion Detection System and Intrusion Prevention System are features that perform deep packet inspection to protect a network from intruders.

☐ Network Switch is a hardware component that allows multiple devices to be connected simultaneously with a router.

☐ Access Point is a hardware device (for example, a wireless router) that allows multiple Wi-Fi devices to connect with the network.

☐ Any device that can connect to a network or Internet has a Media Access Control (MAC) address which provides a device-specific number.

☐ A router can be compromised with unauthorized port forwarding, sending data to a device without your permission. Disable unauthorized port forwarding

☐ A router can be compromised with unauthorized DMZ, sending all traffic to a device that you have not specified. Turn off DMZ.

☐ A router can be compromised with RAM-resident malware. Such malware can be removed by power-cycling the router.

☐ All routers have firmware that requires updating from time to time. See your user manual for instructions.

☐ Bluetooth is a wireless standard very similar to Wi-Fi, but intended for short range, typically under 30 feet.

☐ Bluetooth uses the same unregulated 2.4 GHz frequency range as do older Wi-Fi, microwave ovens, garage door openers, wireless headphones, and wireless landline phones.

☐ Bluetooth devices can be paired with your computer from *System Settings > Bluetooth.* Pair only the devices you recognize.

12 Web Browsing

Distrust and caution are the parents of security.
–Benjamin Franklin[1]

What You Will Learn in This Chapter

- HTTPS good, HTTP not
- Choose a browser
- Configure browser security
- Private browsing
- Secure web searches
- Clear browser history
- Find, remove, and add browser extensions
- Detect fraudulent websites
- Do Not Track and Fingerprinting
- Recover from a web scam
- Use Tor
- Surface web, deep web, and dark web
- Find and fix hacked accounts
- Manage ad targeting information

---

[1] *https://en.wikipedia.org/wiki/Benjamin_Franklin*

What You Will Need in This Chapter

- No additional resources required.

12.1 HTTPS

Due to an extraordinary marketing campaign, everyone knows the catchphrase: *What happens in Vegas, stays in Vegas*. With few exceptions, web surfers think the same thing about their visits to the web.

As of this writing, over 30% of all websites[2] still use the very vulnerable HTTP[3] (Hypertext Transport Protocol) to relay information and requests between user and website and back again. HTTP sends all data in clear text. This means anyone snooping on your network connection anywhere between your computer and the web server can easily see everything you are doing.

The primary alternative to HTTP is HTTPS[4] (Hypertext Transport Protocol Secure). HTTPS accounts for 68% of active websites and uses the SSL[5] (Secure Socket Layer) encryption protocol to ensure all traffic between the user and server is military grade encrypted.

- Note: Due to the continuing efforts of the major web browsers to eradicate insecure sites from the web, at this time, a user of Firefox, Chrome, etc. must set special security exemptions to access data from an insecure site. Also, further restrictions in Google Search Engine Optimization[6] (SEO) guidelines have removed nearly all insecure protocol website listings from their index. This does leave one to wonder why we have not yet reached universal encryption.

---

[2] *https://w3techs.com/technologies/details/ce-httpsdefault*

[3] *https://en.wikipedia.org/wiki/Hypertext_Transfer_Protocol*

[4] *https://en.wikipedia.org/wiki/HTTPS*

[5] *https://en.wikipedia.org/wiki/Transport_Layer_Security*

[6] *https://en.wikipedia.org/wiki/Search_engine_optimization*

Any time you visit a web page that is secured using https, it is reflected in the URL or address field of your web browser, typically as a small lock icon in the URL. It looks like this:

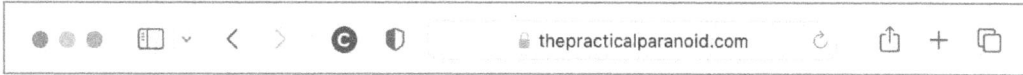

When connected securely to a website, snoops cannot monitor your actions. However, they still can see you are connected to the site. If you want to shield yourself completely, continue reading to our chapter on using a Virtual Private Network (VPN.)

Fortunately, most (but not all) websites that use HTTPS automatically route a browser to the websites' HTTPS even if a browser requests a non-secure HTTP format.

Having to remember to connect only to an HTTPS site for each web page is an impossible task. As we noted, over 30% of websites still do not have an HTTPS option, resulting in an error page and wasted time when you presume to add an S to an HTTP address. But whenever possible you do prefer to work with HTTPS sites because otherwise all your communication is visible to people you do not know/authorize.

There are three options to resolve this:

- Use Safari version 15 or higher. Safari in macOS 12 and higher has a built-in mechanism like HTTPS Everywhere and will automatically router your URL request to the HTTPS version of the website if it is available.

- Automate your browser's attempt to connect to sites that use HTTPS. Firefox and Brave are good choices to configure such automation. Both have *HTTPS Everywhere*[7] built into them. If you cannot use either of these, you may be able to add HTTPS Everywhere as an extension to your preferred browser.

- Encrypt your entire online session using VPN. More on this on the *Internet Activity* chapter.

---

[7] *https://www.eff.org/https-everywhere/faq*

12.2 Choose a Browser

There are many web browsers on the market, with each placing different emphases on various features. The most popular browsers for macOS are Safari, Mozilla Firefox, and Google Chrome. Safari is included with macOS, while Chrome and Firefox are available as free downloads. As this course is all about security, we focus on Safari, Brave, and Tor which are my favorites because of the security features they offer currently.

Why might you want to replace your current browser with another? There may be features–including additional security and privacy–offered by an alternative browser. Check out the features of browsers listed below:

Browser/Base	Platform	Price	Notable Features	Privacy
Brave Chromium	Android, iOS, macOS, Linux, Windows	Free	Built-in HTTPS Everywhere, Malware protection, Ad block	High
Chrome/ Chromium	Android, Chromebook, iOS, Linux, macOS, Windows	Free	We no longer advise using Chrome Browser due to growing security and privacy issues	Poor
Edge Chromium	Windows 11	Free (included with Windows 11)	The new Internet Explorer 11 is still IE	Poor
Firefox Firefox (Gecko)	Android, iOS, Linux, macOS, Windows	Free (Open Source)	Add-ons, Privacy, History and Bookmarks can be shared between your devices running Firefox	High
Opera Chromium	Android iOS, Linux, macOS, Windows	Free	Built in VPN, sends info between devices, ad blocker	High

Browser/Base	Platform	Price	Notable Features	Privacy
Safari **Webkit**	macOS, iOS	Free	Built into Apple ecosystem, synchronized Bookmarks and History, forced HTTPS, Fingerprint blocking, malware protection	High
Tor **The Tor Project** **Firefox (Gecko)**	macOS, Linux, Windows	Free	Anonymous browsing, Built-in HTTPS everywhere, can access the dark web	Very High

12.2.1 Assignment: Secure Browsing with Brave

Brave was designed to be the most secure chromium browser available. It includes anti-malware, ad-blocking, HTTPS Everywhere.

In this assignment, you install the Brave browser to help ensure a more secure browsing experience.

Download Brave

1. Open a web browser, then surf to *https://www.brave.com*

2. Tap the *Download* button.

3. At the *Download Brave* page, select *Download Brave*. Brave downloads to your computer.

4. Double tap to open the *Brave.dmg* file.

5. Drag Brave into your Applications folder.

Configure Brave

6. Open the Brave browser.

7. Select the *Brave* menu > *Preferences*.

8. From the sidebar, select *Get Started*. My settings are shown below:

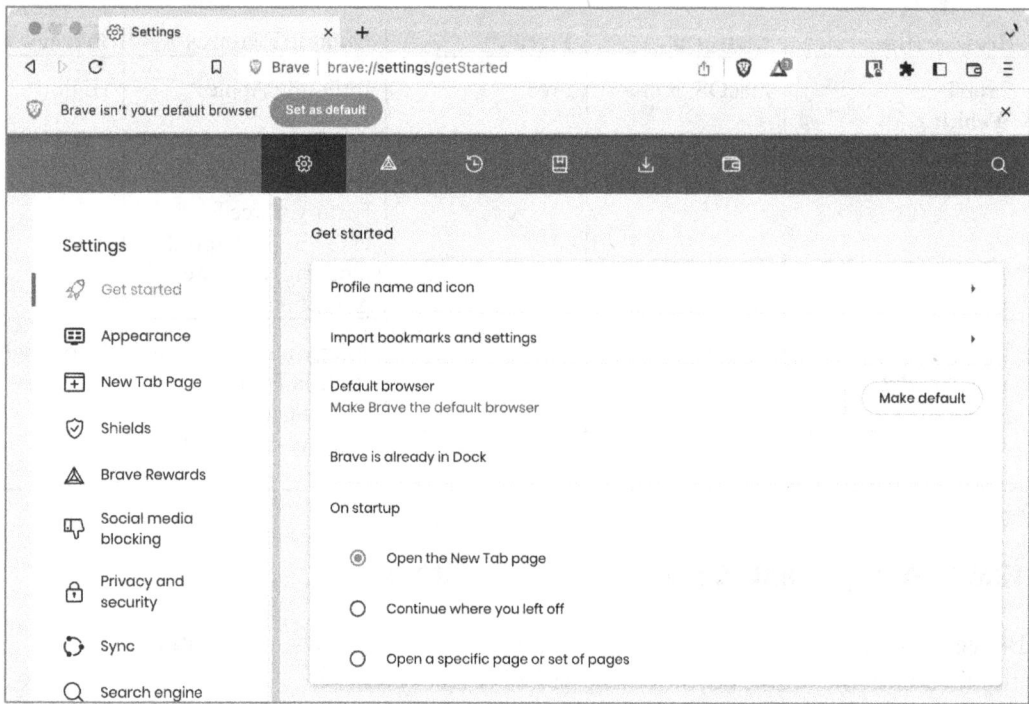

9.   In the main area of the page select *Import bookmarks and settings*.

10. Select what data and settings you want to import, then select the *Import* button.

11. If prompted with the *Full Disk Access required* alert, follow the on-screen instructions. Once complete, repeat the previous step.

12. In the sidebar select *Appearance.* Configure to your taste. My settings are shown below:

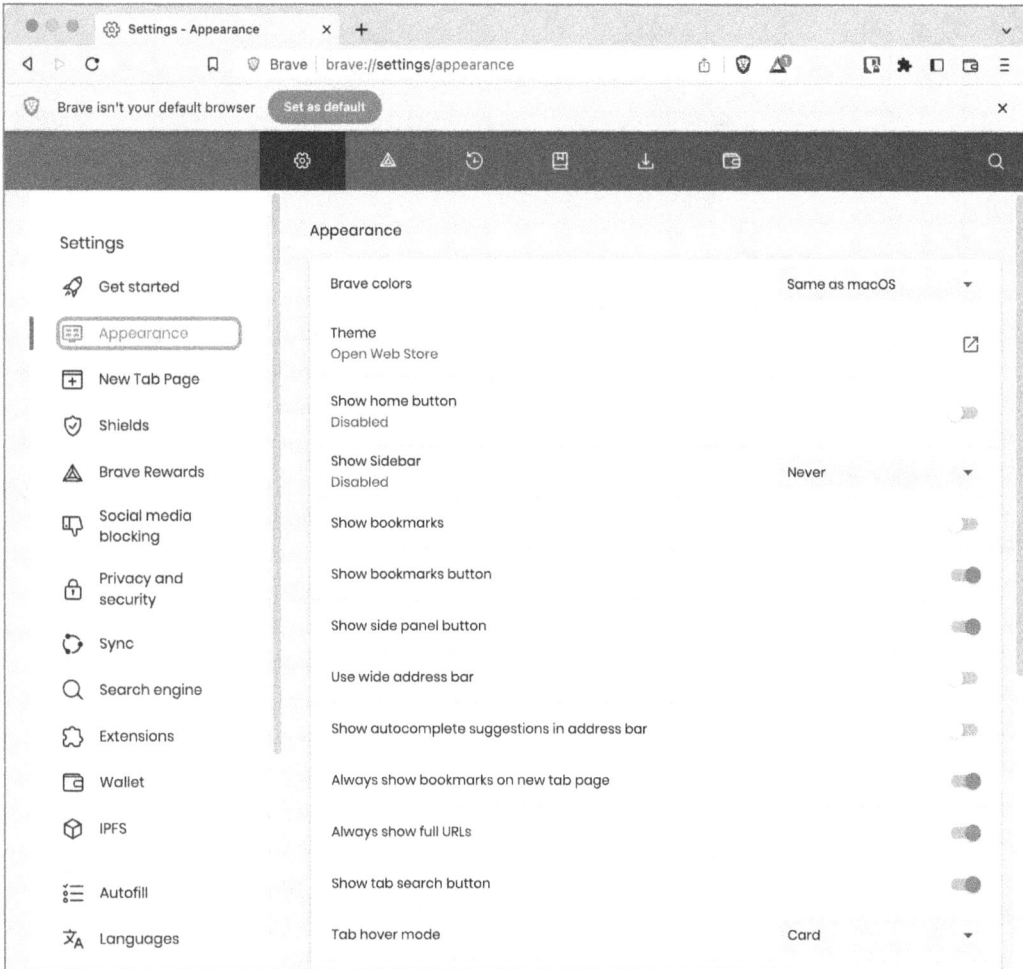

13. In the sidebar select *New Tab Page.* Configure to your taste.

14. In the sidebar select *Shields*. Configure to your taste. My recommended settings are shown below:

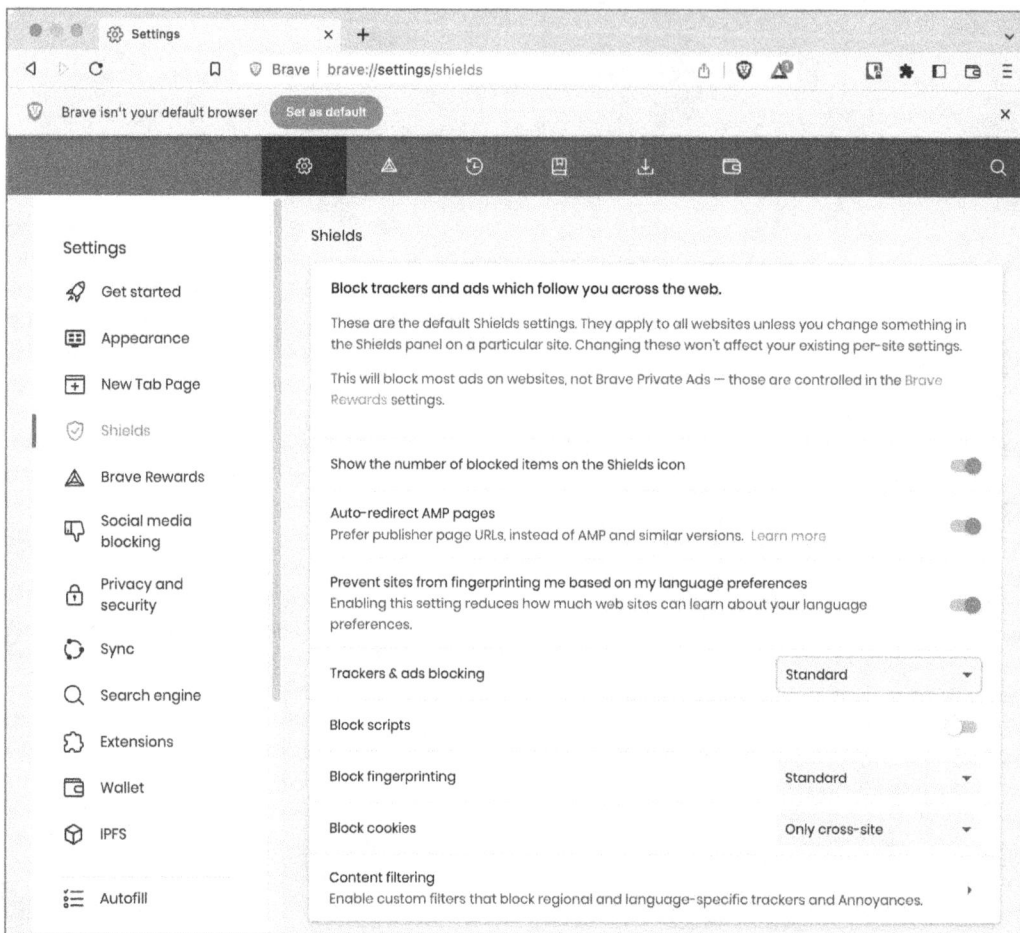

15. In the sidebar select *Social media blocking.* Configure to your taste. My settings are shown below:

- NOTE: Although I have *Allow Google login buttons on third party sites* and *Allow Facebook logins and embedded posts* enabled, never use Google or Facebook logins for other sites. This allows Google and Facebook access to your data on those sites.

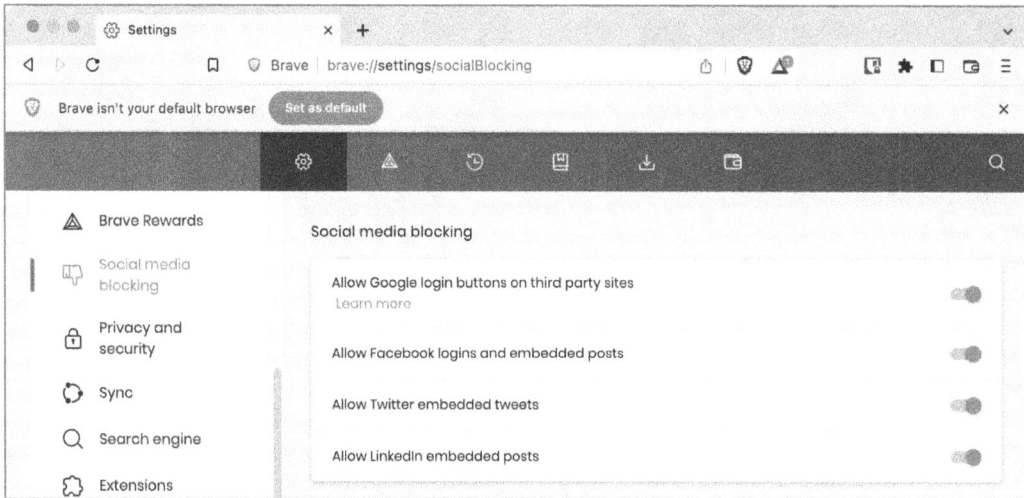

16. In the sidebar select *Privacy & Security*. Configure to your taste. My recommended settings are shown below:

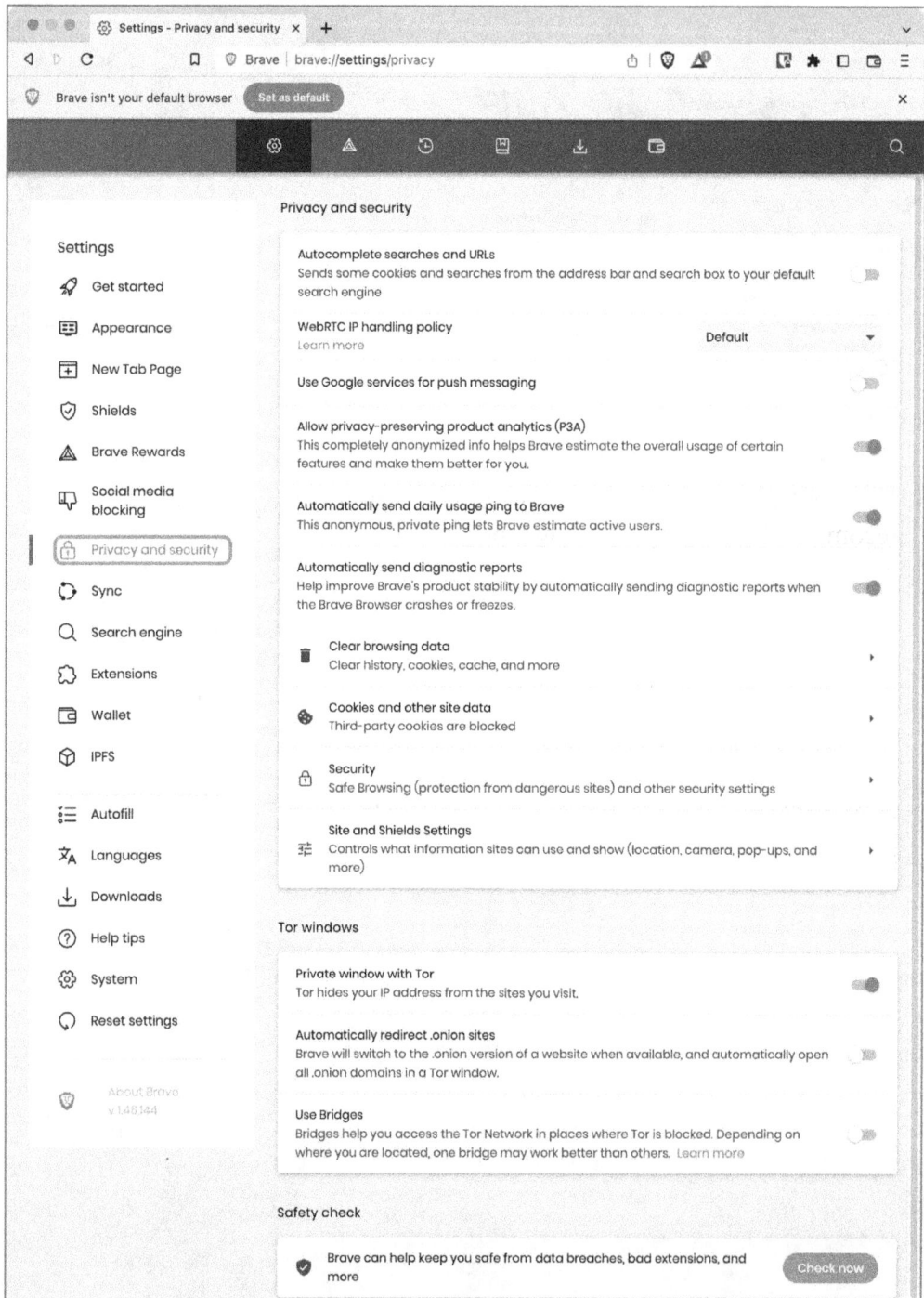

Settings - Privacy and security ✕ +

◁ ▷ C 🔖 🛡 Brave | brave://settings/privacy 🖆 🛡 🗛⁰ 🖥 ✱ ▢ 🗒 ☰

🛡 Brave isn't your default browser (Set as default) ✕

⚙ ⚠ 🕒 💾 ⬇ 🗒 🔍

Privacy and security

Autocomplete searches and URLs
Sends some cookies and searches from the address bar and search box to your default search engine

WebRTC IP handling policy
Learn more

Default ▼

Use Google services for push messaging

Allow privacy-preserving product analytics (P3A)
This completely anonymized info helps Brave estimate the overall usage of certain features and make them better for you.

Automatically send daily usage ping to Brave
This anonymous, private ping lets Brave estimate active users.

Automatically send diagnostic reports
Help improve Brave's product stability by automatically sending diagnostic reports when the Brave Browser crashes or freezes.

🗑 **Clear browsing data**
Clear history, cookies, cache, and more ▸

🍪 **Cookies and other site data**
Third-party cookies are blocked ▸

🔒 **Security**
Safe Browsing (protection from dangerous sites) and other security settings ▸

⛭ **Site and Shields Settings**
Controls what information sites can use and show (location, camera, pop-ups, and more) ▸

Settings

🚀 Get started

⊞ Appearance

⊞ New Tab Page

🛡 Shields

⚠ Brave Rewards

👎 Social media blocking

🔒 Privacy and security

⟳ Sync

🔍 Search engine

🧩 Extensions

🗃 Wallet

◈ IPFS

≔ Autofill

🅰 Languages

⬇ Downloads

? Help tips

⚙ System

○ Reset settings

🛡 About Brave
v 1.48.144

Tor windows

Private window with Tor
Tor hides your IP address from the sites you visit.

Automatically redirect .onion sites
Brave will switch to the .onion version of a website when available, and automatically open all .onion domains in a Tor window.

Use Bridges
Bridges help you access the Tor Network in places where Tor is blocked. Depending on where you are located, one bridge may work better than others. Learn more

Safety check

🛡 Brave can help keep you safe from data breaches, bad extensions, and more (Check now)

17. In the sidebar select *Sync.* If you wish to have your Brave browser settings and bookmarks synchronized between your devices, follow the on-screen instructions. Else, go to next step.

18. In the sidebar select *Search engine*.

19. Set *Search engine used in the address bar* to *DuckDuckGo*.

20. In the sidebar select *Extensions.* Configure to your taste. My settings are shown below:

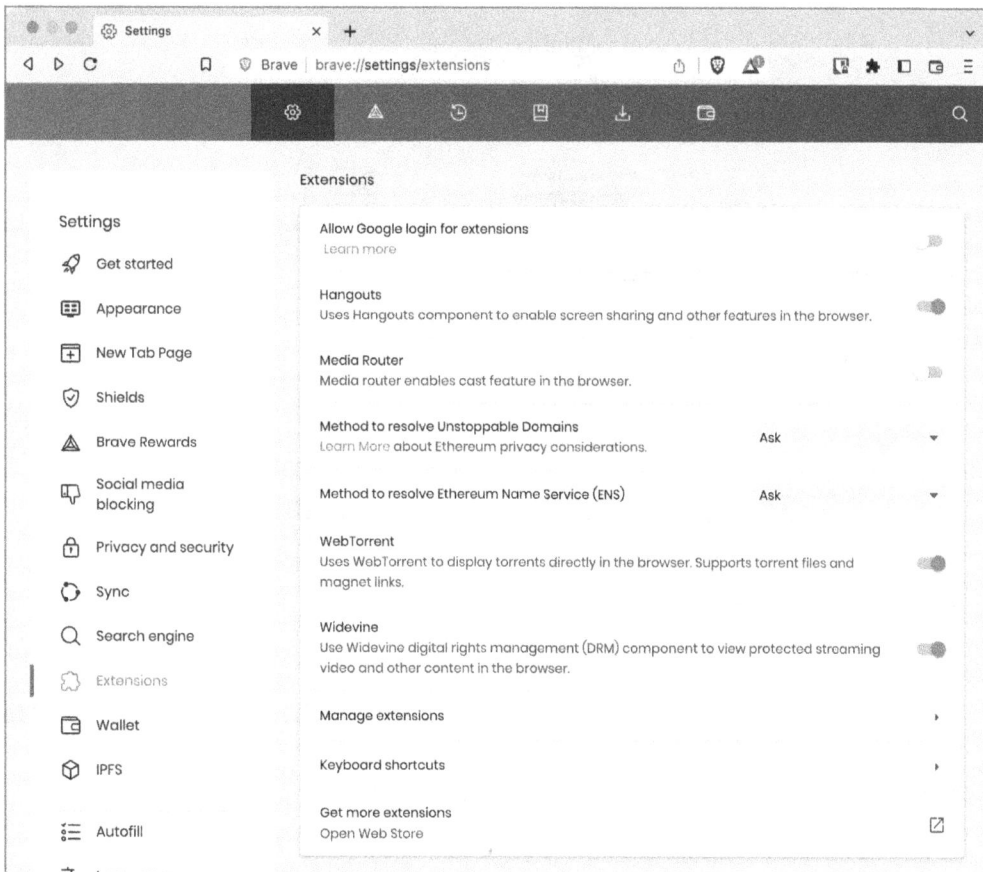

21. In the sidebar expand *Additional Settings > System*.

22. Enable *Use hardware acceleration when available*.

23. When complete, close *Settings*.

Customize for each website

Brave can be customized on a site-by-site basis.

24. With Brave open, visit a website (in this example, *thepracticalparanoid.com*).

25. The Brave Shield in the right side of the address bar displays how many trackers have been blocked.

26. Tap the *Brave Shield*. The *Shield configuration* window opens.

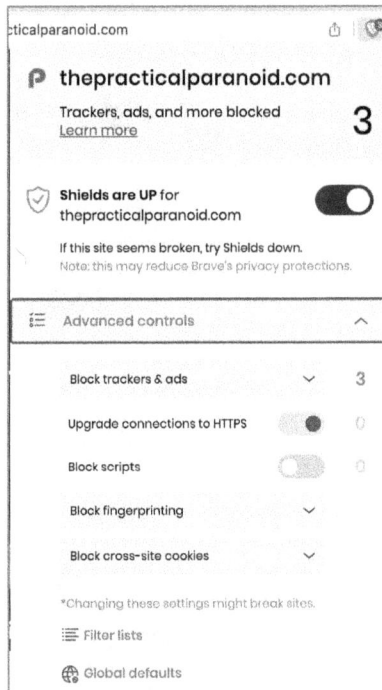

27. If a website is behaving poorly due to the default Brave settings, this is where you can create a custom setting for this site.

28. Quit Brave.

12.2.2 Assignment: Secure Browsing with Safari

In this assignment, you secure Safari.

1. Open Safari, then select the *Safari* menu > *Settings*.

2. Tap the *General* tab. Configure the *General* tab to your taste. My settings are shown below:

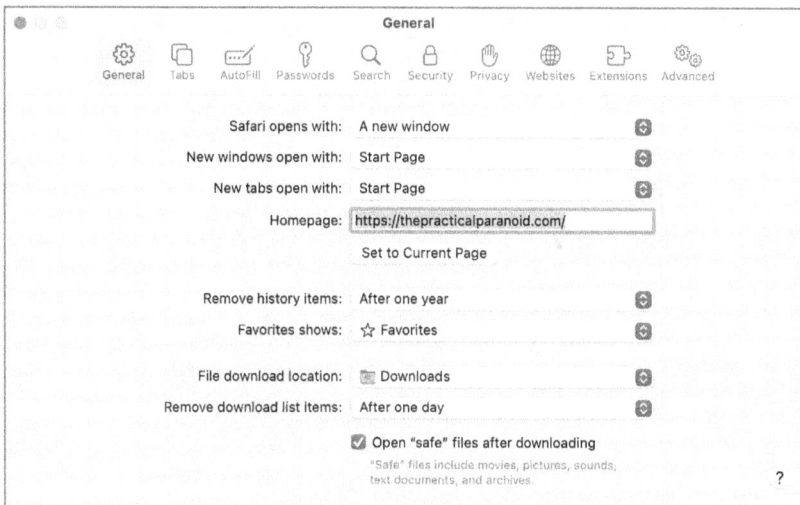

3. Tap the *Tabs* tab. Configure to your taste. My settings are below:

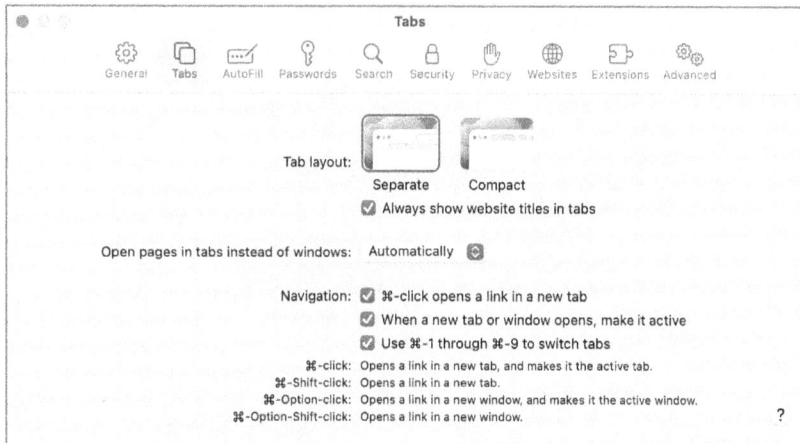

4. Tap the *AutoFill* tab. Configure to your taste. My settings are below:

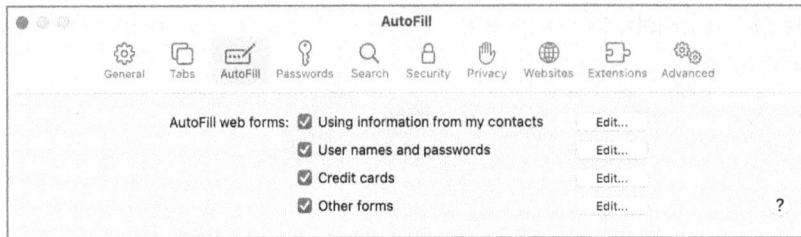

5. There is nothing to configure in the *Passwords* tab.

6. Tap the *Search* tab. Configure as below:

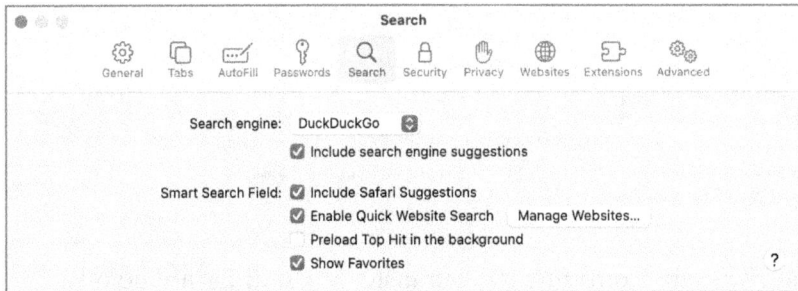

7. Tap the *Security* tab. Configure as below:

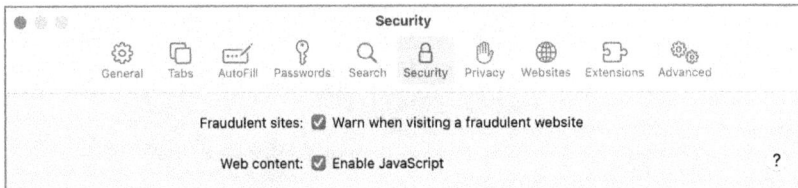

8. Tap the *Privacy* tab. Configure as below:

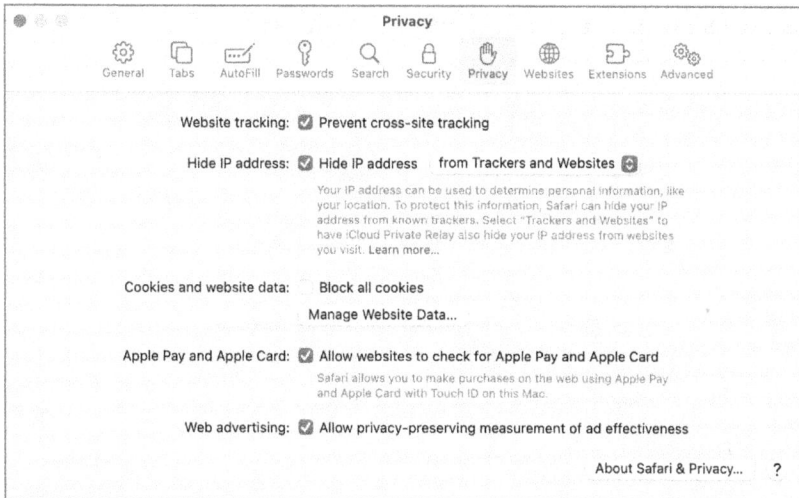

- Note: Introduced with macOS 12 is the option to *Hide IP address*.

9. Tap the *Websites* tab. Select each item in the side bar one by one to verify no sites have unwanted access.

10. Tap the *Extensions* tab. Verify only desired extensions are installed and enabled.

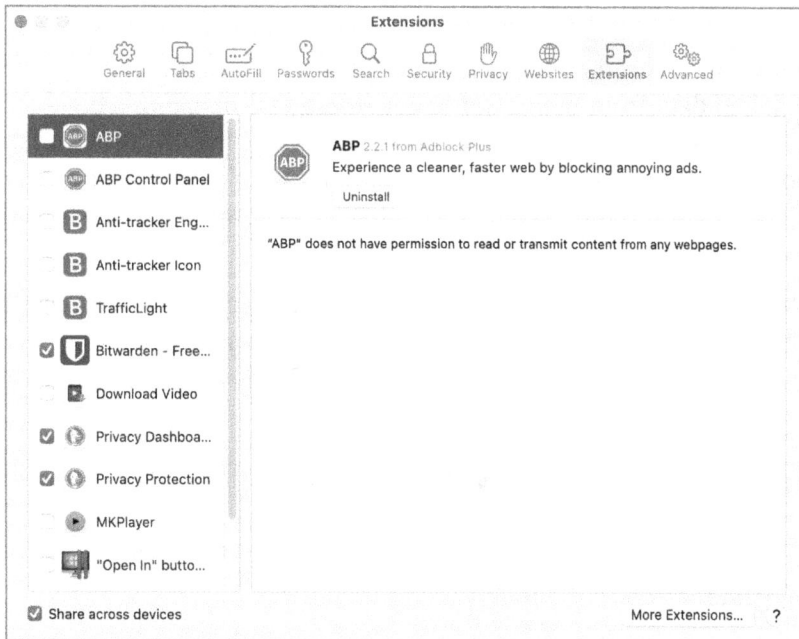

11. Tap the *Advanced* tab. Configure to your taste. My settings are below:

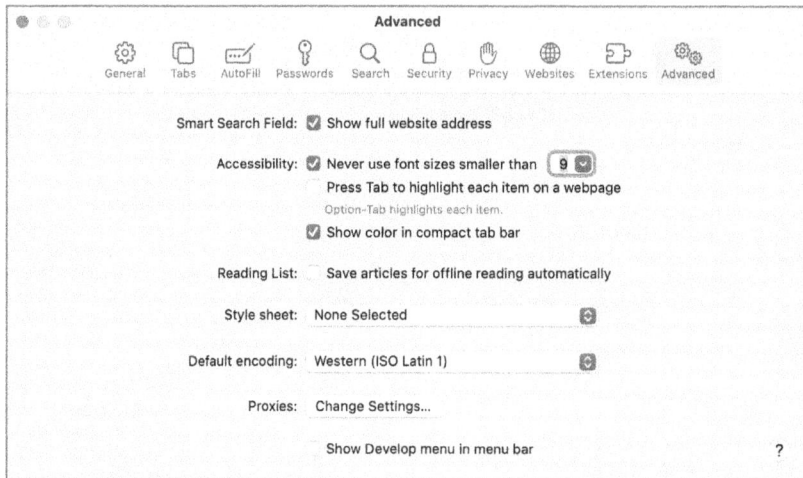

12. Close Safari Preferences

12.3 Private Browsing

Private Mode (Safari) and *Private Window* (Brave) are features that prevent any normally cached data from being written to storage while using a browser. This data includes browsing history, passwords, usernames, list of downloads, cookies, and cached files. Private browsing is an essential tool if you work on a computer where your account is shared (what's with that?), or if there is the possibility that someone else may examine your browsing habits.

- Note: This does not prevent your company IT department or Internet Provider from seeing or recording your browsing habits. To accomplish this, you must be using VPN (Virtual Private Network) on your device. More on that later.

12.3.1 Assignment: Brave Private Window

In this assignment, you enable Brave Private Window browsing.

1. Launch Brave.

2. Select *File* menu > *New Private Window.* A new Private Window opens.

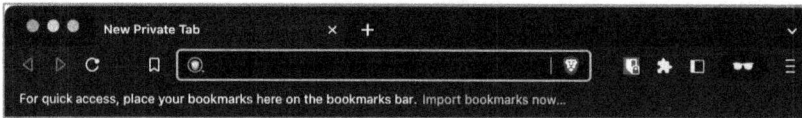

3. The indicators of being a Private Window are the URL field purple highlight, *New Private Tab* as the tab name, and the *Private* eyeglasses to the far right of the URL field.

12.3.2 Assignment: Safari Private Browsing

In this assignment, you enable Private Browsing within Safari.

1. From the Safari *File* menu, select *New Private Window.*

2. A new Safari window appears. You can see *Private Browsing* is enabled by the *Address* field being dark.

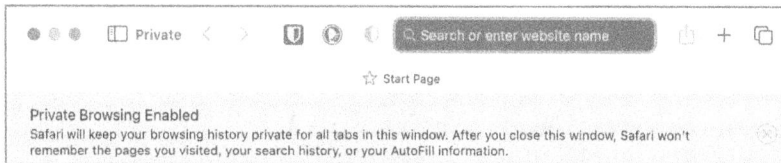

12.4 Secure Web Searches

When you perform a search, the search criteria and sites visited may be collected and stored by the browser developer, search engine, your Internet Service Provider, and the trackers associated with sites visited. The cookies assigned from one website can communicate with other sites and webpages you open, allowing both the sites and advertisers to maintain logs of your activities and building a profile of your search history so that your search results are unique and tailored to your interests. Thus, if you query "gun," you can get ads related to this topic and beyond, such as militia websites. If you query baby animals, you can see postings more related to that, such as videos of kittens saved by dogs. The point is, not only are your habits and interests noted, but they are perpetuated by algorithms designed to send you to political and marketing bots from unknown parties. These sources can give different people different information, both true and false, and thereby manipulate our knowledge bases. It is no wonder that cultural division can result.

Not so with the *DuckDuckGo* search engine. DuckDuckGo's policy is to keep no information on user searches. Subsequently, your search activity is private, and all search results are identical for everyone. All users see the same data when searching on the same terms.

Browsers offer the option to make DuckDuckGo your default search engine. This is a big step towards providing a better level of privacy on the Web.

12.4.1 Assignment: DuckDuckGo for Brave Search

In this assignment, you change the default Brave search engine to the secure DuckDuckGo.

1. Open Brave.

2. Select *Brave > Preferences.*

3. From the sidebar, select *Search Engine.*

4. From the *Search engine used in the address bar* pop-up menu, select *DuckDuckGo.*

5. Exit from *Preferences.*

12.4.2 Assignment: DuckDuckGo For Safari Search

In this assignment you change the default Safari search engine from Google to the secure search engine DuckDuckGo.

1. Open Safari.

2. Open the *Safari* menu > *Preferences*.

3. Select the *Search* icon from the Toolbar.

4. From the *Search Engine* pop-up menu, select *DuckDuckGo.*

5. Close the Preferences window.

12.5 Clear History

By default, every browser maintains a full history of every site you have visited (when you are not in *Private* mode). Should someone gain access to your device, they can view your browsing history.

You can erase your entire browsing history in one tap. There is no recovery from clearing the browser history.

12.5.1 Assignment: Clear the Brave History

In this assignment, you clear your Brave browsing history.

- **Warning**: there is no recovery from this action. If you wish to keep your history, pass on this assignment.

1. Open Brave.

2. Select *History* menu > *Show Full History.*

3. In the left sidebar, select *Clear Browsing Data.* The *Clear browsing data* window opens.

- Note: If you cannot see the sidebar, expand the width of the Brave page.

4. In the *Clear browsing data* window, select the *Advanced* tab.

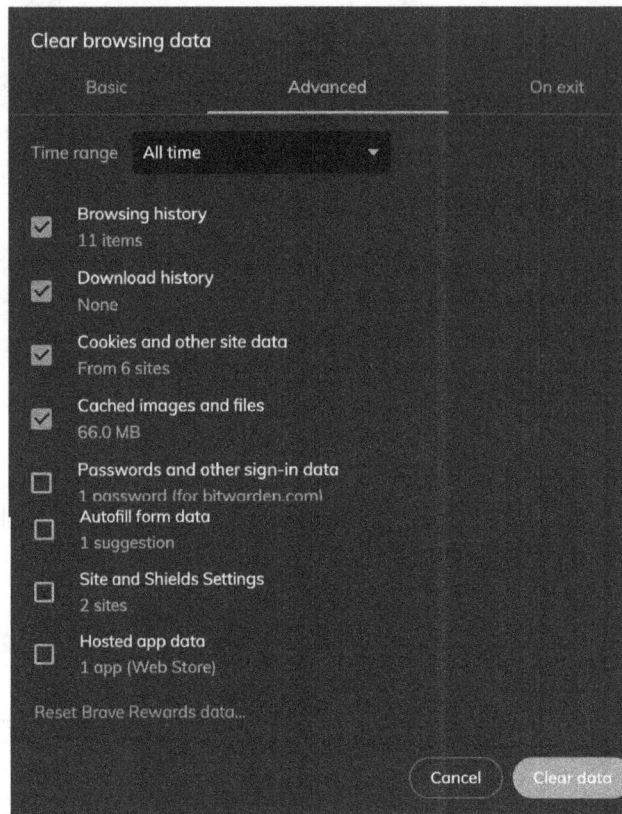

5. From the *Time range* pop-up menu, select the desired range (usually *All time*).

6. Select the items to be cleared, then tap *Clear data* button.

12.5.2 Assignment: Clear Safari History

In this assignment, you clear Safari browsing history.

- **Warning**: there is no recovery from this action. If you wish to keep your history, pass on this assignment.

1. Open Safari, then select the *History* menu > *Clear History…*

2. A dialog box opens asking for what time frame of your history that you wish to clear. Make your selection (usually *all history*), then select the *Clear History* button.

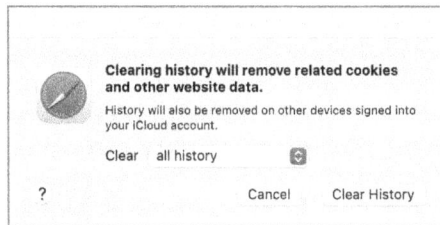

The Safari history is now cleared as you defined.

12.6 Browser Extensions

One of the great advances in personal computer software development was the concept of plug-ins or extensions[8]. These small strings of code add functionality to the host application. In the case of web browsers, a function may be anything from the ability to encrypt web-based email to the ability to view proprietary video formats.

The bad news about extensions is that they run with the full power of the host application. This means that a malicious extension may have the power to secretly redirect your web browser to fake websites (such as a phony copy of your bank website), or harvest all your passwords, monitor your purchases, etc.

There are many malicious extensions. It is vital to install only those that you need to install, to know which are installed, and to rid yourself of unnecessary extensions.

12.6.1 Assignment: Find, Remove, and Add Brave Extensions

In this assignment, you see the Brave extensions, determine if they are what you need, remove those not needed, and search for potential additions.

1. Open Brave.
2. Select the *3-Line* menu in the right top corner > *Extensions.* All currently installed extensions display.

---

[8] *https://en.wikipedia.org/wiki/Plug-in_(computing)*

3. If you see any extensions you do not remember installing, perform an Internet search to discover what they do, and if they present a vulnerability.

4. If you determine you do not want an Extension installed, select the target extension, then select the *Remove* button under the target extension.

5. To find potential additional extensions, select the *Web Store* link located under the installed extensions. The *Chrome Web Store* opens to display all Chrome extensions. As Brave is based on the *Chromium Project,* it can use all Chrome Browser Extensions.

Add the Bitwarden Extension

6. While in the Chrome Web Store, search for, then download/install *Bitwarden*.

7. Once installed, select the *Manage Extensions* icon at the far right of the *Address bar > Pin*. This forces the Bitwarden Brave extension to display in the Address bar.

8. Open the Bitwarden Brave extension, link to your Bitwarden account created in the Password chapter, then configure to your taste.

12.6.2 Assignment: Find, Remove, and Add Safari Extensions

In this assignment, you see the installed Safari extensions, determine if they are what you need, remove those not needed, and search for potential additions.

1. Open Safari.

2. Select the *Safari* menu > *Preferences* > *Extensions* tab.

3. All currently installed extensions display in the sidebar.

4. If you see any extensions you do not remember installing, perform an Internet search to discover what they do and if they present a vulnerability.

5. If you determine you do not want an extension installed, select that target extension in the sidebar, then select the *Uninstall* button under the target extension in the main body area.

6. To find potential additional extensions, select the *More Extensions* button at the bottom right of the *Extensions* tab window. The Apple App Store opens to display all Safari extensions.

12.7 Fraudulent Websites

As of this writing, there are almost 2,000,000,000 active websites[9]. Within that, there may be millions (hundreds of millions?) of fraudulent websites. Of the diverse types of fraud found on the Internet[10], among the most common are websites that misrepresent who they are. This may be in the form of appearing like Bank of America, but with a URL of perhaps https://bankofamerica.cm, instead of the true https://bankofamerica.com. In this case, the *https* means the site is encrypted, not necessarily a valid or safe site. A criminal is hoping for someone to accidentally type `.cm` instead of `.com`. Once at the fake site, you would enter your account and password as usual. The difference is that this time, a criminal now obtains your credentials–and all your money within minutes.

As a side note, in this specific example at the time of this writing (for a much earlier version of MacOS), this URL is a scam site. But not for the scheme mentioned. When I went to *http://bankofamerica.cm*, I was routed to the following:

```
VIRUS FOUND

A website you visited today has infected your Mac with a virus.

Press OK to begin the repair process.

                                                              Close
```

If we look at the full URL, it is: http://apple.com-----systemmessenger.com/dgkg/?city=Albuquerque&region=New%20Mexico&country=US&ip=71.222.135.33&isp=Qwest%20Communications%20Company%20Llc&os=OS%20X&osv=OS%20X%2010.11%20El%20Capitan&browser=Safari&browserversion=Safari%209&voluumdata=BASE64dmlkLi4wMDAwMDAwNi01Yzg0LTRjNjYtODAwMC0wMDAwMDAwMDAwMDBfX3ZwaWQuLmRl...

From this URL, we can see that the criminal site attempts to appear as though it is Apple reporting that I have a virus. They have also discerned my city, state, IP

---

[9] *http://www.internetlivestats.com/total-number-of-websites/*
[10] *https://en.wikipedia.org/wiki/Internet_fraud*

address, Internet provider (Qwest Communications), and that I am using OS X 10.11 El Capitan with Safari version 9.

If I were the typical user, I'd think there was a virus present and tap the *OK* button as recommended. You also may have noticed the criminal was bright enough to do all of this, but not bright enough to put an *OK* button in the script!

So, I tap the *Close* button. I'm presented with a new window:

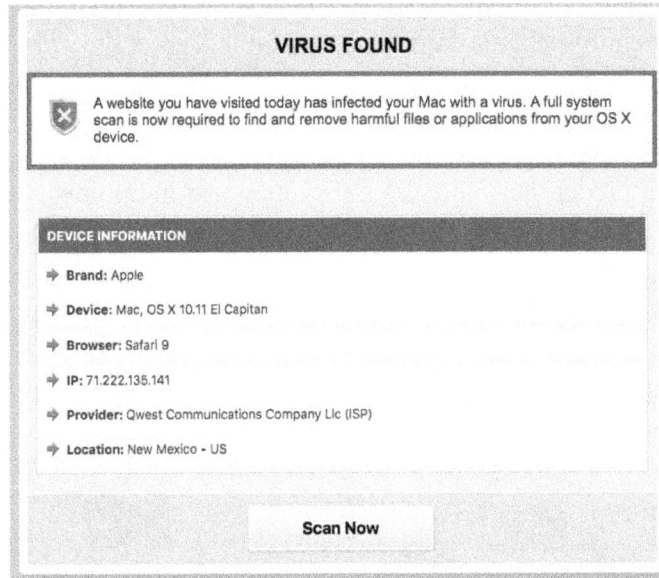

Hey, who needs an *OK* button when the *Close* button provides the intended scam! So, let's see what happens when I tap the *Scan Now* button:

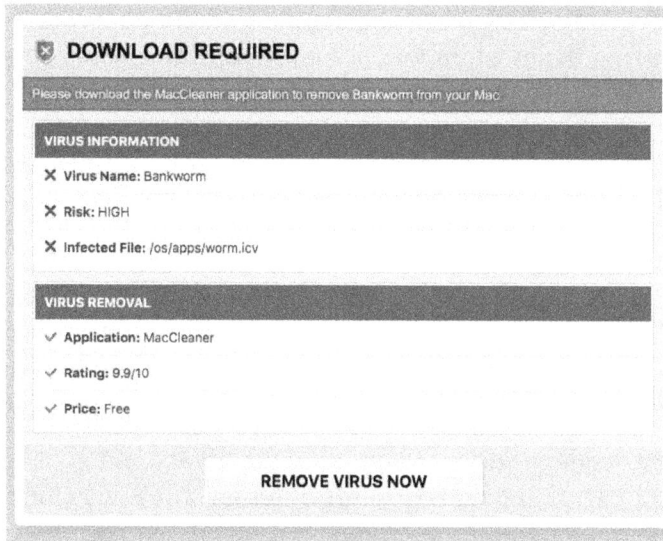

It appears the scammer thinks I am infected with the *Bankworm* virus, and the infected file is *in /os/apps/worm.icv*. The only real problem I see is that there is no such file, and no such directory.

But they are offering a free solution to my non-existent problem. Let's see where that takes us by tapping the *Remove Virus Now* button:

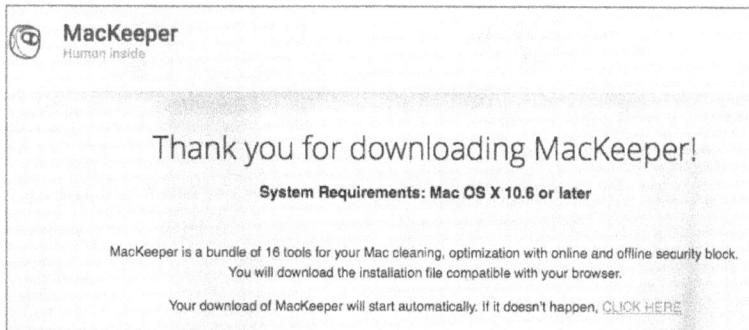

MacKeeper? Really! This product lost a class action lawsuit for deceptively advertising its functionality[11].

If you have followed along so far, just trash the MacKeeper download.

---

[11] *https://topclassactions.com/lawsuit-settlements/closed-settlements/94767-mackeeper-class-action-settlement/*

So, how to protect yourself against fraudulent sites? We go through the few steps that can be taken, but the most important tool is your awareness.

12.8 Do Not Track and Fingerprinting

Most websites track which pages you visit, how long you stay on each page, and other metrics to better understand their visitors. That is a little creepy. Imagine going to the library and having a librarian look over your shoulder as you scan the card catalogue and record each of the books and pages you glance at.

Now let's take the analogy further. You leave the library and go across town to have lunch. You look around and the same librarian still is watching and recording everything you looked at on the menu and what you ate.

Later you go on a date, and the librarian is sitting right behind you in the theater, noting who you are with, what scenes you react to, and more.

Web browsing isn't much different—except the snoop is normally invisible in the form of cookies, trackers, and browser fingerprinting.

Any website can initiate cookies on your browser. These keep a record of the pages you visit on the site. But they have evolved to report all the other places you visit and things that you do. Therefore, you can visit Amazon, look up my books, quit the web browser, launch it, go to Facebook, and see an ad for my books!

In addition, there are third-party trackers that are integrated with many sites. When you visit one site, you are uniquely identified. In this way, when you visit another site, the tracker knows who you are. Trackers can piece together a solid psychological, sociological, and financial profile on you.

There is the option to disable cookies, but most of your websites demand cookies be enabled to visit the site. And there are many ways to track you, even without cookies. To get a good sense of the philosophy of the trackers, look at this slightly dated *Guide to Cookieless Tracking.*[12]

---

[12] *https://www.ionos.com/digitalguide/online-marketing/web-analytics/browser-fingerprints-tracking-without-cookies/*

Trackers are invisible by default. However, we do have tools to thwart them. Some web browsers have a preference setting to ask web sites not to track you. Notice the term is *ask*. Good luck with that. Of course, you can turn off cookies but often that means the site will refuse to deal with you. You must decide how much you need the product or information offered by the site.

Device fingerprinting[13] is a bit more of a problem. When opening a webpage, it is possible for the site (or a search engine) to uniquely identify your device amongst the hundreds of millions of other devices on the internet. Once you have been fingerprinted, it is easy to track your activity across the internet. As ghastly as this is, there is a fix to make your fingerprint appear more generic.

There are promises from browser developers to build in fingerprinting blocking technology. But it is currently only reasonably implemented in Safari, Tor, and Brave. There are browser extensions to block fingerprinting as well.

Safari calls their fingerprinting blocking *Intelligent Tracking Prevention*. It includes the following features:

- Blocks third-party cookies, preventing advertisers and websites from following you around the internet through cross-site tracking.

- Deletion of third-party cookies and website data unless you have directly interacted with the content provider.

- Blocks web tracking through machine learning.

- Provides a substitute IP address, hiding your actual IP address.

The only 100% solution at this time is to use a bootable thumb drive with Tails[14] installed, then do nothing to alter the drive or OS setup. When booting from Tails, you look exactly like all other Tails users, and it is impossible to fingerprint you.

---

[13] *https://en.wikipedia.org/wiki/Device_fingerprint*
[14] *https://tails.boum.org/*

12.8.1 Assignment: View Your Device Fingerprint

Each device on the internet is unique based on its combination of features, traits, fonts, settings, etc. This can be used to identify the device, then to track it across the internet. This is called a *device fingerprint*[15].

In this assignment, you view your device fingerprint.

View Fingerprint with Amiunique.org

1. Open a browser, then go to *https://amiunique.org.*

2. Select *View my browser fingerprint.* In a minute or two a summary of your unique fingerprint is displayed.

3. Scroll down to see all your fingerprint attributes.

View Fingerprint with Panopticlick.eff.org

4. Open a new browser window, then go to *https://panopticlick.eff.org.* This site is similar in function to *amiunique.org* and provides some unique information.

5. Enable the *Test with a real tracking company* checkbox.

6. Select the *Test Your Browse* button. After a few minutes, a results summary appears.

7. Scroll down to the *Detailed Results* area > *Select a characteristic* pop-up menu. Then tap *Detailed View* to display to view all your fingerprint details.

12.9 Web Scams

Over the past couple of years, a new type of scam has become popular. Instead of directly compromising the user computer, web sites are compromised or deliberately designed to be malicious.

When you visit such a site, you may receive a pop-up window stating something to like: Your computer has been found to be infected with XX viruses. Please call Apple at XXX-XXX-XXXX to have this infection removed.

---

[15] *https://en.wikipedia.org/wiki/Device_fingerprint*

Upon calling the provided toll-free phone number (which, of course, is not Apple, but the scammer), with your permission, they install remote control software. After looking around your computer, they assure you they can remove the malware for only $$$. Just recite to them your credit card information.

There are two big problems here. First, a criminal installed remote control software that allows the criminal access any time they wish. They can access your usernames, passwords, banking, and your other information. The second is that they now have your credit card information.

Do you think you are too smart, educated, or street-smart to fall victim? These attributes may put you at *greater* risk![16, 17]

12.9.1 Assignment: Recovering from a Web Scam

In this assignment, you examine your browsers for possible modifications. What to do if a scam happens to you?

Reset Homepage setting

1. Do not call the number offered by the scammer! These are criminals.

2. In most cases, the malicious website has modified your web browser preferences to make the malicious page your home page so that is the page you see when you turn on your computer. Open your browser *preferences* (in this example, Safari) > *General* > *Homepage* field. If the *Homepage* field is not what you had set, delete, or reset the entry.

---

[16] *https://www.marketplace.org/2011/09/30/why-even-smart-people-fall-scams/*
[17] *https://assets.aarp.org/rgcenter/econ/fraud-victims-11.pdf*

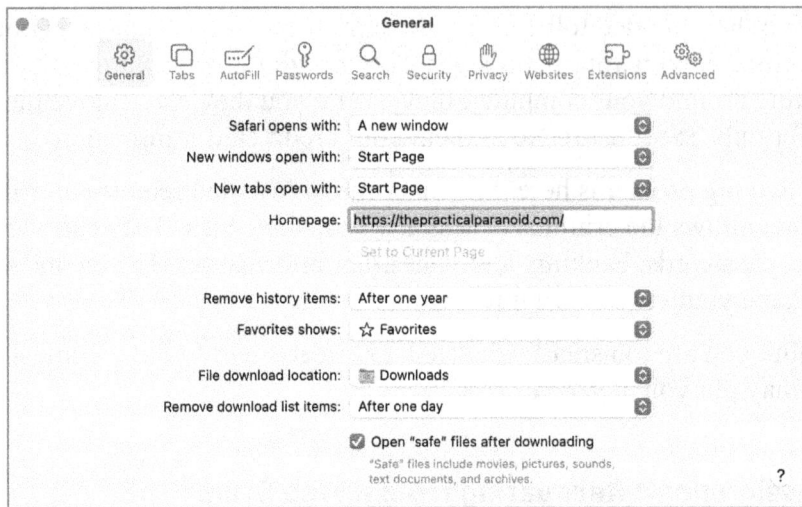

3. Malicious attacks on a browser often block access to your browser preferences. If you cannot access your browser preferences, open *System Settings > Desktop & dock,* then enable *Close windows when quitting an app.*

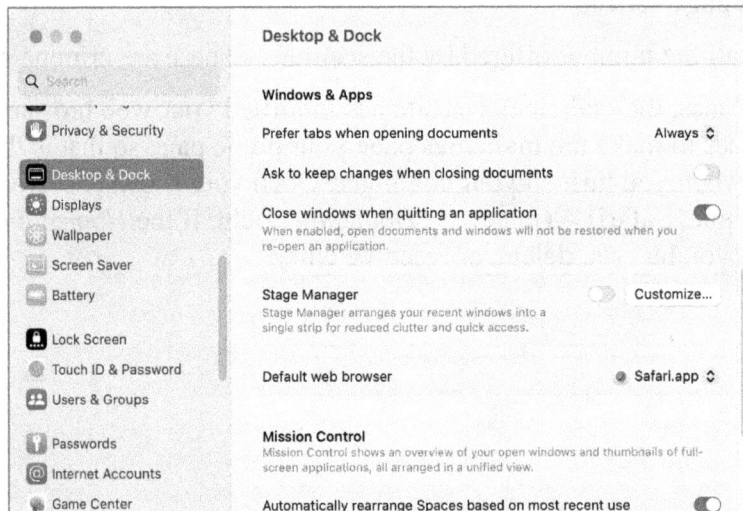

4. Quit Safari.

5. Open Safari to test. You should no longer have the malicious page open.

6. Repeat steps 1-5 for each of your browsers.

Check for DNS Changes

Some malicious sites change your DNS settings. By doing so, they can redirect your browsing to fake sites they control but that look like the real deal. Their hope is you will enter sensitive data they can use such as passwords and bank account numbers.

7. Open *System Settings > Network*.

8. Authenticate to unlock.

9. Select the *Advanced* button.

10. Select the *DNS* tab.

11. In the *DNS Servers* field, verify the correct server address(es) are listed. If not, edit them to the correct address(es).

- Note: By default, network devices take on the DNS server addresses fed to them by the network router. In turn, by default, routers take on the DNS server addresses fed to them from their Internet provider. I recommend using the Cloudflare DNS server, with addresses of 1.1.1.1 and 1.0.0.1. These servers do not track your activities and are currently the fastest available. This can be configured in your router portal, as well as from within the *System Settings > Network >* Tap target network *Details* button *> DNS* tab.

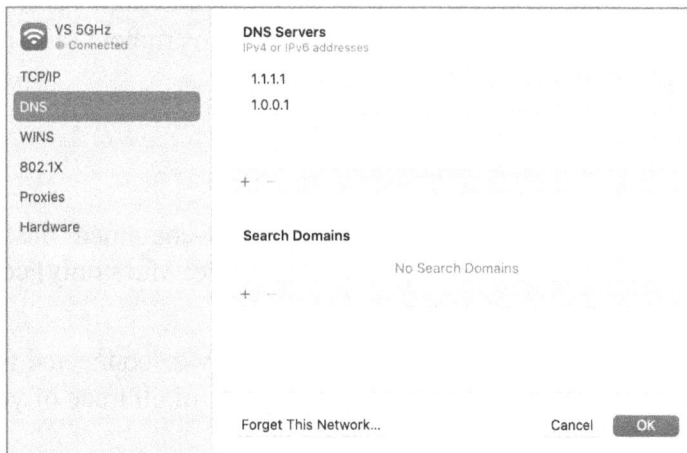

12. Select the *OK* button.

13. Select the *Apply* button.

14. Quit System Settings.

Done!

12.10 Tor

Tor[18] is a technology developed by the US Department of the Navy that enables anonymous web browsing. Anonymous browsing is accomplished by first encrypting your data destined for the internet, then routing it through multiple *onion* routers that make up the *onion network[19]*. Each router only knows which router the data packets came from, and which router to forward to, so as your data exits the final router it is not possible to trace the packets back to their origins.

Anonymous browsing has long since been released to the open-source community for the public to use in the form of the *Tor Browser*. The Tor browser is a stripped-down browser with the unique ability to communicate with the onion network.

Many people in the security community are strong supporters of Tor, including Edward Snowden. Entire books have been written on just Tor. I'm not so sadistic as to subject you to that. However, keep in mind that although Tor may keep your Internet activities anonymous to criminals, the US government (and likely other governments) can compromise Tor.

Tor advantages:

- Strong anonymity for all activity on the Internet.

- Can be used with Tails[20] which is a bootable, self-contained, flash drive. Tails can run on most Windows, Linux, and Apple (Intel Macs only) computers and it leaves behind no trace of activity.

- The bootable Tails flash drive can be immediately disconnected from the host computer, causing the computer to erase memory of all trace of your session, and reboot.

Tor disadvantages:

---

[18] *http://en.wikipedia.org/wiki/Tor_(anonymity_network)*
[19] *https://en.wikipedia.org/wiki/Onion_routing*
[20] *https://tails.boum.org*

- Tor was developed by the US Department of the Navy. It is possible there are back doors only the government knows about.

- The US government has been forthright about having its own Tor relays in place, which enable it to monitor online activity. Not a big deal if you only wish to be anonymous to criminals. It is a big deal if you wish to be anonymous while performing black-market deals for Aunt Rose's raisin Noodle Koogle recipe.

- The onion network and Tor help ensure your anonymity by bouncing your web traffic around at least three random onion routers. This can dramatically reduce your browsing performance.

These features make Tor ideal for those in oppressed countries, journalists working undercover, and anyone who may need to use someone else's computer and leave no trace behind.

Tor works by encrypting your packets as they leave your computer, routing the packets to a Tor relay computer hosted by thousands of volunteers on their own systems, many of which are co-located at ISPs. The relay knows where the packet came from, and the next relay the packet is handed to, but that is all. The user computer automatically configures encrypted connections through the relays. Packets pass through several relays before being delivered to the intended destination. Tor uses the same relays for around 10 minutes, then different relays are randomly selected to create the next path for 10 minutes.

Alas, there is no free lunch. The encryption process and the relay process combine to create *latency*, which means a delay in processing. Most users experience around a four-fold performance degradation. So, if accessing a web page without Tor normally takes 3 seconds, it may take 12 seconds with Tor.

Even though Tor does as good a job as anything to keep you anonymous on the Internet, you must take precautions to protect your identity. These steps include:

- Do not enable JavaScript when using Tor. JavaScript has been used to track users within the Tor network.

- Do not reveal your name or other personal information in web forms.

- Do not customize the Tails boot flash drive. This creates a unique digital fingerprint that can be used to identify you.

- Connect to sites that use HTTPS, so your communications are encrypted point to point.

Tor by itself is at best a partial solution. It can protect your anonymity while surfing the web. At the very least, this still leaves email (if you are not using web-based email) and messaging to be secured. A bigger issue is what to do when you need to use a computer and leave no trace behind on that system. This is where Tails comes into play.

Tails is a Linux Debian fork designed with two primary purposes:

- Provide a highly secure operating system in a format that can be booted from either DVD or thumb drive on almost any PC or Apple computer, and

- Include the tools and applications necessary to provide a secure, anonymous Internet experience.

What this means is that, with Tails, you can create a thumb drive that has an operating system capable of booting almost any computer, whereby you can run Tor for secure anonymous Internet activity, send and receive email that is securely encrypted with GPG/PGP, and message with others in complete privacy. Then, when you remove the Tails thumb drive, there is absolutely no record of your activity on either the computer *or* the thumb drive.

For those of you chomping at the bit to just use Tor, we start there. When your curiosity has been satisfied, please take the next step to learn Tails[21].

12.10.1 Assignment: Use Tor for Anonymous Internet Browsing

Tor is a stripped down, simplified web browser, providing the highest level of privacy while browsing. The trade-off is lack of flexibility, as modifying Tor's preferences or adding extensions makes your identity unique.

In this assignment, you download and install Tor.

1. As a first step, we need to know our public IP address. This information is used a few steps away to verify Tor has hidden our address. Open a web browser to *https://whatismyipaddress.com*. Leave this window open for reference.

---

[21] *https://tails.boum.org*

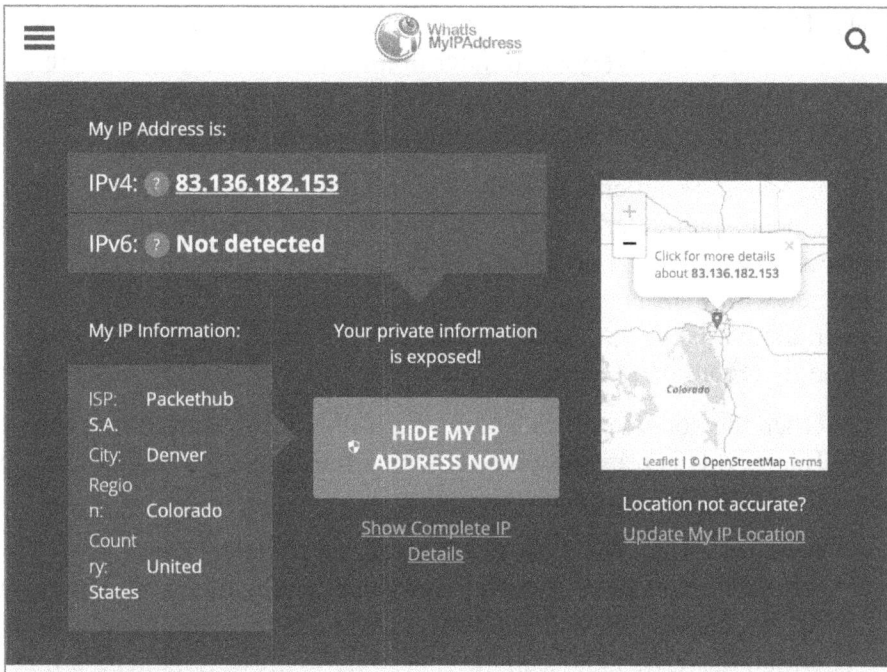

2. Open a web browser then go to *https://www.torproject.org > Download Tor Browser*.

3. Select the *Download Tor Browser* button for your OS. The Tor installer begins to download.

4. While the download is in progress, note these other steps that one must take to ensure privacy is maintained. These include:

   • Use the Tor Browser. If you are concerned about protecting your privacy and security, do not use other browsers.

   • Do not torrent over Tor. If you wish to file-share via torrent, do not use Tor. Torrent is painfully slow, it slows down others using the Tor network, and in many cases, torrent software bypasses all the security and anonymity precautions built into Tor and is broadcast unencrypted.

   • Do not enable or install browser plugins in Tor. Tor is designed to protect your security and anonymity. Many innocuous-looking plugins break that security.

- Use HTTPS versions of websites. Tor has *HTTPS Everywhere* built in (more on HTTPS Everywhere later in this book). It forces a secure connection if a website has an option for https. This enables a point-to-point encryption between your computer and the web server.

- Do not open documents downloaded through Tor while online. Many documents–particularly .doc, .xls, .ppt, and .pdf–contain links, macros, or resources that force a download when the document is opened. If they are opened while Tor is open, they reveal your true IP address, and you lose your anonymity and security. If you are concerned about these issues, we strongly recommend doing this instead:

  o Open the documents on a computer fully disconnected from the Internet. This prevents any malicious files from "phoning home" or infecting your computer.

  o Install a Virtual Machine (VM) such as Parallels, Fusion, or VirtualBox (as of this writing, only Parallels works with Apple silicon Macs), configured with no network connection, and open documents within the VM. This is an alternate way to prevent malicious files from phoning home or infecting your computer.

  o Or use Tor while within Tails. This is an alternative way to prevent malicious files from phoning home or infecting your computer.

- Use bridges and/or find company. Tor cannot prevent someone from looking at your Internet traffic to discover you are using Tor, even if they cannot determine what it is you are doing. If this is a concern for you, reduce the risk by configuring Tor to use a *Tor Bridge relay* instead of a direct connection to the Tor network. Another option is to have many other users running Tor on the same network. In this way, your use of Tor is hidden.

5. Open the Tor installer. It mounts and opens a disk image onto the Desktop.

6. Drag the TorBrowser.app into your *Applications* folder.

7. Open Tor.

8. It is vital to test your connection to verify your IP address is hidden/changed. While in Tor, return to *https://whatismyipaddress.com*. You should have a different IP address than you did outside of tor.

Wahoo! You are now on Tor, completely anonymous and encrypted on the Internet. Your next assignment is to configure Tor.

12.10.2 Assignment: Configure Tor Preferences

One of the first things one should do when launching an application for the first time is to configure its preferences. No different for Tor.

In this assignment, you configure Tor preferences.

- Prerequisite: Completion of the previous assignment, or having Tor installed on your device.

1. Open TorBrowser, then select the *3-Line* menu (top right) > *Settings* > *General*. Configure to your taste. My settings are shown below:

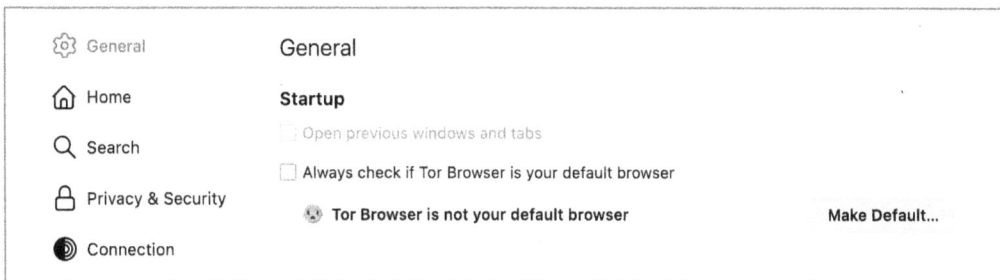

Startup

☐ Open previous windows and tabs

☐ Always check if Tor Browser is your default browser

    🌑 **Tor Browser is not your default browser**   Make Default...

Tabs

☑ Ctrl+Tab cycles through tabs in recently used order

☑ Open links in tabs instead of new windows

☑ When you open a link, image or media in a new tab, switch to it immediately

☐ Confirm before closing multiple tabs

☑ Confirm before quitting with ⌘Q

Language and Appearance

Website appearance

Some websites adapt their color scheme based on your preferences. Choose which color scheme you'd like to use for those sites.

◉ Tor Browser theme　　　　○ System theme　　　　○ Light　　　　○ Dark

Manage Tor Browser themes in Extensions & Themes

General

🏠 Home

🔍 Search

🔒 Privacy & Security

◉ Connection

🧩 Extensions & Themes

Colors

Override Tor Browser's default colors for text, website backgrounds, and links.

Manage Colors...

Fonts

Default font   Default (Times)   ⌄   Size   16   ⌄   Advanced...

Zoom

Default zoom   **100%**   ⌄

☐ Zoom text only

Language

Choose the languages used to display menus, messages, and notifications from Tor Browser.

English (en-US)   ⌄   Set Alternatives...

Choose your preferred language for displaying pages   Choose...

☑ Check your spelling as you type

Files and Applications

Downloads

Save files to
☐ Downloads   **Choose...**

☐ Always ask you where to save files

Applications

Choose how Tor Browser handles the files you download from the web or the applications you use while browsing.

🔍 Search file types or applications

Content Type	▲	Action
◈ AV1 Image File (AVIF)		⬤ Open in Tor Browser
▣ Extensible Markup Language (XML)		🖼 Save File
🗐 irc		🗐 Always ask
🗐 ircs		🗐 Always ask
🗐 mailto		🗐 Always ask
Portable Document Format (PDF)		⬤ Open in Tor Browser
▣ Scalable Vector Graphics (SVG)		🖼 Save File
◈ WebP Image		⬤ Open in Tor Browser

What should Tor Browser do with other files?

⦿ Save files

◯ Ask whether to open or save files

General
🏠 Home
🔍 Search
🔒 Privacy & Security
◉ Connection

🔌 Extensions & Themes
❓ Tor Browser Support

Tor Browser Updates

General

Home

Search

Privacy & Security

Connection

Tor Browser Updates

Keep Tor Browser up to date for the best performance, stability, and security.

Version 102.6.0esr (64-bit)

Show Update History...

☺ Tor Browser is up to date

Check for updates

Allow Tor Browser to

○ Automatically install updates (recommended)

○ Check for updates but let you choose to install them

Performance

☑ Use recommended performance settings Learn more

These settings are tailored to your computer's hardware and operating system.

Browsing

☑ Use autoscrolling

☑ Use smooth scrolling

☑ Always use the cursor keys to navigate within pages

☐ Search for text when you start typing

☑ Control media via keyboard, headset, or virtual interface Learn more

☐ Recommend extensions as you browse Learn more

Extensions & Themes

☑ Recommend features as you browse Learn more

Tor Browser Support

2. Select *Home* from the sidebar, then configure to your taste.

3. Select *Search* from the sidebar. Configure *Default Search Engine* to *DuckDuckGo*. Configure other Search settings to your taste.

4. Select *Privacy & Security* from the sidebar. My settings are shown below:

General

Home

Search

Privacy & Security

Connection

Onion Services

Prioritize .onion sites when known. Learn more...

◉ Always

◯ Ask every time

Cookies and Site Data

Your stored cookies, site data, and cache are currently using 0 bytes of disk space. Learn more

ⓘ In permanent private browsing mode, cookies and site data will always be cleared when Tor Browser is closed.

☑ Delete cookies and site data when Tor Browser is closed

Clear Data...

Manage Data...

Manage Exceptions...

Logins and Passwords

☐ Ask to save logins and passwords for websites

   ☐ Autofill logins and passwords

   ☐ Suggest and generate strong passwords

   ☑ Show alerts about passwords for breached websites Learn more

☐ Use a Primary Password Learn more

Formerly known as Master Password

Exceptions...

Saved Logins...

Change Primary Password...

Onion Services Authentication

Some onion services require that you identify yourself with a key (a kind of password) before you can access them. Learn more

Saved Keys...

Extensions & Themes

History

Tor Browser will   **Use custom settings for history**   ⌄

☑ Always use private browsing mode   **Clear History...**

☐ Remember browsing and download history

☐ Remember search and form history

☐ Clear history when Tor Browser closes   Settings...

Address Bar

When using the address bar, suggest

☐ Browsing history

☐ Bookmarks

☐ Open tabs

☐ Shortcuts

☐ Search engines

Change preferences for search engine suggestions

Permissions

⊙ Location   Settings...

▢ Camera   Settings...

⏺ Microphone   Settings...

⊟ Notifications   Learn more   Settings...

⏵ Autoplay   Settings...

▭ Virtual Reality   Settings...

☑ Block pop-up windows   Exceptions...

☑ Warn you when websites try to install add-ons   Exceptions...

General
Home
Search
Privacy & Security
Connection

Security

Security Level

Disable certain web features that can be used to attack your security and anonymity.
Learn more

○ **Standard**

All browser and website features are enabled.

◉ **Safer**   Custom

Disables website features that are often dangerous, causing some sites to lose
functionality. Restore Defaults
- JavaScript is disabled on non-HTTPS sites.
- Some fonts and math symbols are disabled.
- Audio and video (HTML5 media), and WebGL are click-to-play.

○ **Safest**

Only allows website features required for static sites and basic services. These
changes affect images, media, and scripts.

Deceptive Content and Dangerous Software Protection

☑ Block dangerous and deceptive content Learn more

　☑ Block dangerous downloads

　☑ Warn you about unwanted and uncommon software

General

Home

Search

Privacy & Security

Connection

Deceptive Content and Dangerous Software Protection

☑ Block dangerous and deceptive content   Learn more

   ☑ Block dangerous downloads

   ☑ Warn you about unwanted and uncommon software

Certificates

☑ Query OCSP responder servers to confirm the current validity of certificates

View Certificates...

Security Devices...

HTTPS-Only Mode

HTTPS provides a secure, encrypted connection between Tor Browser and the websites you visit. Most websites support HTTPS, and if HTTPS-Only Mode is enabled, then Tor Browser will upgrade all connections to HTTPS.

Learn more

◉ Enable HTTPS-Only Mode in all windows

○ Enable HTTPS-Only Mode in private windows only

○ Don't enable HTTPS-Only Mode

Manage Exceptions...

Extensions & Themes

Tor Browser Support

5. Select *Connection* from the sidebar. Configure to your taste. My settings are below:

Connection

Tor Browser routes your traffic over the Tor Network, run by thousands of volunteers around the world.   Learn more

⊕ **Internet:**   Test   ◉ **Tor Network:** ✓ Connected

Quickstart

Quickstart connects Tor Browser to the Tor Network automatically when launched, based on your last used connection settings.

☑ Always connect automatically

Bridges

Bridges help you access the Tor Network in places where Tor is blocked. Depending on where you are, one bridge may work better than another.   Learn more

Add a New Bridge

Choose from one of Tor Browser's built-in bridges   **Select a Built-In Bridge...**

Request a bridge from torproject.org   **Request a Bridge...**

Enter a bridge address you already know   **Add a Bridge Manually...**

Advanced

Configure how Tor Browser connects to the internet   **Settings...**

View the Tor logs   **View Logs...**

General
Home
Search
Privacy & Security
Connection
Extensions & Themes
Tor Browser Support

6. I advise you to not install extensions and themes. This restraint keeps your Tor fingerprint anonymous and avoids potential malicious software installation.

7. Close *Settings*.

Great work! You are now ready to use Tor to browse the Internet securely and anonymously.

But remember, Tor is just one small part of *real* anonymity and security on the Internet. Many in the Internet Security field believe that to do this right, you want a bootable Tails thumb drive. Learn all about it in the Tails[22] home page.

---

[22] *https://tails.boum.org*

12.10.3 Assignment: Browse with Tor

In this assignment, you browse a site with both its www and onion address.

● Prerequisite: Installation of Tor.

1. Open Tor browser.

2. In the address field, enter `https://torproject.org`. This is the home for the Tor developers.

3. If your Tor preferences are configured as recommended (*Privacy & Security > Browser Privacy > Always*), Tor will automatically resolve the site to the available .onion network.

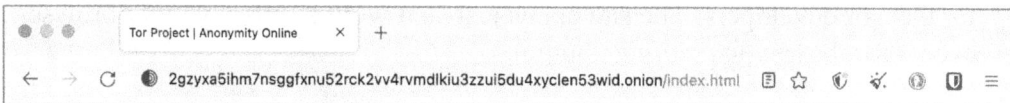

4. Note the address change.

5. Quit Tor.

12.10.4 Assignment: Brave Private Window with Tor

With its recent upgrade, the Brave browser now includes the ability to encrypt internet traffic over the onion network. This allows us to use a very modern browser, include extensions, *and* maintain security and privacy while browsing.

If you require the highest level of privacy while browsing, Tor is your best option. If you can trade a little privacy protection for much greater flexibility, Brave is a great solution. In this assignment, you use Brave to browse with Tor.

● Prerequisite: Installation of the Brave browser.

1. As a first step, you need to know your public IP address so you can use this information a few steps away to verify that Tor has hidden your address. Open Brave to *https://whatismyipaddress.com*. Write down *Your IP*.

2. Select the *3-Line* menu > *New private window with Tor*.

3. The page changes, stating that you are now in a *Private Window with Tor*. The difference between this and the normal Brave (or another browser) window is:

- You can access websites on the onion network. These are recognizable because their URL ends with .onion, for example *cakewalk.onion*.

- All your browsing, on the surface web as well as the onion network, is encrypted and bounced between routers to hide your identity.

- Due to the encryption and bouncing, your web access is significantly slower.

4. In this private window, go to *https://whatismyipaddress.com.* The verification that you now are using Tor is that your IP address has not only changed, but this is now an entirely different network:

5. In the address field, enter `https://torproject.org`. This is the home for the Tor developers. The site opens just as it would with any other browser, except for the *.onion available* button:

6. Note the address shown in the address field.

7. Tap the *Tor* button. This instructs Brave open a new private window, then connect with the sites .onion site, using the .onion network:

8. Note the address change.

9. Quit Brave.

12.11 Surface Web, Deep Web, and Dark Web

The internet can be viewed as having three layers: *Surface, Deep, and Dark.*

The *Surface Web* is anything that can be indexed by a typical search engine like Google. A good way to visualize this is to go to a web site–for this example, Amazon.com–then tap any links found on the site. This is how a search engine works. It follows links.

The *Deep Web*[23] is anything that cannot be indexed by a typical search engine like Google. A good way to visualize this is to go back to Amazon.com, then find the price of a specific product from 15 years ago–but you are limited to doing so by only using links. It cannot be done. Such information could be found using the search feature within Amazon, but that isn't how a search engine works. It can only use links. This product information would be considered Deep Web content.

The *Dark Web*[24] is that portion of the Internet content that requires specific software, configurations, or authorization to access. The Dark Web is a small part of the Deep Web. The most common tools used to access the Dark Web include Tor[25], I2P[26], and Freenet[27].

Tor not only allows you to have anonymous access to your regular web sites, but it is also one of the few tools offering a gateway to the Dark Web. Web sites on the dark web are also called *Onion sites,* as they end with *.onion.*

Although the dark web is primarily thought of as a collection of sites to sell illegal products and services, there are also good and responsible uses for it. For example, in repressive countries such sites provide an avenue for freedom workers and reporters to securely exchange information with sources (Ed Snowden did this), and there are sites to provide resources for whistleblowers.

As the dark web is not indexed by Google, Bing, or any other standard search engine, how do you go about discovering its resources? The list is in constant flux. One of the better ways of finding dark web directory listings is a web search of *dark web directory.*

12.12 Hacked Accounts

Do you know if you've been pwned?

---

[23] *https://en.wikipedia.org/wiki/Deep_web*
[24] *https://en.wikipedia.org/wiki/Dark_web*
[25] *https://en.wikipedia.org/wiki/Tor_(anonymity_network)*
[26] *https://en.wikipedia.org/wiki/I2P*
[27] *https://en.wikipedia.org/wiki/Freenet*

"WHAT!?!" is probably the first thing that just went through your mind. No, it's not a typo. *Pwn*, as defined in the dictionary, is to be totally defeated or dominated. Although most used when trouncing your online game opponent, it is also used to describe when your email or online accounts have been hacked.

Unfortunately, there is a pretty good chance that you have been pwned!

Several websites track email and online account breaches. My favorite is *haveibeenpwned.com*.

12.12.1 Assignment: Find Your Hacked Accounts

In this assignment, you search the haveibeenpwned.com database to discover if any of your online accounts have been hacked/pwned.

1. Open a web browser to *https://haveibeenpwned.com*. The home page appears.

2. Enter your email address, then tap the *pwned?* button.

3. In a few seconds, the results display.

4. Make a note of the sites with breaches. In most cases, your concern should focus on breaches that include account name or email address, and password.

- Note: the site may report there have been *pastes*. A paste is information about your pwned account that has been "pasted" to a surface web site such as Pastebin.com. These are services favored by hackers.

5. Close your browser.

12.12.2 Assignment: What to Do Now That You Have Been Hacked

In this assignment, you take action to repair any found breaches.

- Prerequisite: Completion of the previous assignment.

1. If the breach is a specific website, open a web browser to the breach site.

2. Change your account password, following these best practices:

- Passphrase is a minimum of 15 characters in an easy-to-enter phrase.

- Use each password/passphrase for only one site. Should a site become compromised, and your password harvested, many automated hacking systems then use your credentials at every bank, online store, etc. they know of to see if you, like most folks, use one password for everything.

- Keep a secure record of your passwords/passphrases. I use Bitwarden as my password manager. You can also use a current version of Excel to create an encrypted spreadsheet, which you must update whenever you change your access information.

- Only enter a username and password when operating in a secure web page (https).

3. Repeat steps 1–2 for each breached site.

4. If the breach is not of a specific website or service, there is nothing you can do. In most of these cases, massive collections of breaches from unknown sources are placed on the dark web for sale, or just given away. To repeat: it is so important to not use the same password for more than one site or service.

12.13 Ad Targeting Information

As you browse the Internet, many of the websites you visit track your activity. This information can then be knitted together with the tracking from all your online activity to create a valuable package about who you are, your likes, and your private life.

Although it may not be possible to eliminate all traces of this tracking activity, you do have the power to significantly reduce it.

12.13.1 Assignment: Apple Advertising

Apple provides some transparency into its use of ad targeting information when you visit Apple sites such as App Store, Apple News, and Stocks. For a full disclosure on Apple advertising and privacy policies, visit *https://support.apple.com/en-qa/HT205223*.

In this assignment, you reduce your ad targeting information available to Apple.

1. Open *System Settings* > *Privacy & Security* > *Apple Advertising*.

2. Tap the *About Apple Advertising & Privacy* button. Read the policy document. Then tap *OK*.

3. Tap the *View Ad Targeting Information* button. If Apple has collected information about you from its sites, the information is listed here. Then tap *OK*.

4. Disable *Personalized Ads* checkbox. This does not reduce the number of ads you see on Apple sites, but it does stop Apple from tracking your activity for advertising purposes.

5. Exit *System Settings*.

12.14 Web Browsing Lessons Learned

☐ Websites using http are not secure or private. All data between your device and the server is in clear text.

☐ Websites using https are secure and private. All data between your device and the server is encrypted. The *Lock* icon denotes https.

☐ Browser choice is critical to security and privacy. Some browsers exist to harvest your browsing habits.

☐ Private browsing does not prevent your router, business, or internet provider from recording your browsing activities.

☐ Most search engines are funded by selling your browsing activity records.

☐ Clearing your browser history prevents knowledge of your browsing activities if someone has access to your account on your device.

☐ Browser extensions can monitor your browser activities.

☐ Your browsing activities can be monitored and followed from site to site, from browser to browser and even device to device, based on your device fingerprint.

☐ The internet can be viewed as having three layers. The surface web is anything that can be indexed by a typical search engine. The deep web is anything that cannot be indexed by a typical search engine. The dark web are services and servers accessible only using the secure onion network and onion network compatible browsers, such as Tor and Brave.

☐ It is recommended to review haveibeenpwned.com monthly in case your internet accounts have been compromised.

☐ You can reduce the amount of tracking done by Apple by configuring *System Settings > Security & Privacy > Privacy* tab *> Apple Advertising.*

☐ Brave is one of the most secure and private browsers available.

☐ Brave (desktop version only) has the feature of private browsing with Tor, without needing to use and set up the Tor browser.

☐ Tor may be the most secure and private browser. It uses both encryption and bouncing from router to router to secure your data and shield your identity.

13 Email

Human beings the world over need freedom and security that they may be able to realize their full potential.

–Aung San Suu Kyi[1], Burmese opposition leader and chairperson of the National League for Democracy in Burma

What You Will Learn in This Chapter

- Avoid Phishing
- Use email encryption protocols
- Use TLS encryption
- Configure forced TLS with Paubox
- Use HTTPS encryption
- Use email aliases
- Create end-to-end secure email with ProtonMail
- About PGP/GPG and S/MIME
- Configure SPF, DKIM, and DMARC records
- Use email 2-Factor Authentication

What You Will Need in This Chapter

- [Optional] Administrator access to your private domain control panel and DNS settings.
- [Optional] Email account on private domain.

---

[1] *https://en.wikipedia.org/wiki/Aung_San_Suu_Kyi*

- [Optional] Possess a browser based Outlook.com email account.

13.1 The Killer App

It can be rightfully argued that email is the killer app that brought the Internet out of the geek world of university and military usage and into our homes. Most email users live in some foggy surreal world with the belief they have a God-given or constitutionally given right to privacy in their email communications.

No such right exists. Google, Yahoo!, Microsoft, Comcast, or whoever hosts your email service is likely to turn over all records of your email whenever a government agency asks for that data. In many cases, your email is sent and received in clear text so that anyone along the dozens of routers and servers between you and the other person can clearly read your messages. Add to this knowledge recent revelations about PRISM[2] (2007), also known as SIGAD US-984XN, where the government does not have to ask your provider for records, and you realize the government simply *has permission to access* your records.

Let's talk about what you can do to minimize your loss of privacy.

13.2 Phishing

Phishing[3] is epidemic on the Internet. Phishing is the attempt to acquire your sensitive information by appearing as a trustworthy source. Phishing is most often attempted via email.

The way the process often works is that you receive an email from what appears to be a trustworthy source, such as your bank. The email provides some motivator to contact the source, along with what appears to be a legitimate link to the source website.

---

[2] *https://en.wikipedia.org/wiki/PRISM_(surveillance_program)*
[3] *https://en.wikipedia.org/wiki/Phishing*

When you tap the link, you are taken to what appears to be the trustworthy source (perhaps the website of your bank), where you are prompted to enter your username and password. If you do this, they have you. The site is a fraud, and you have just given the criminals credentials to access your bank account. In a few moments, your account may be emptied.

The key to preventing a successful phishing attack is to be aware of the *real* URL behind the link provided in the email.

The link that appears in an email may have nothing at all to do with where the link takes you. To see the *real* link, hover (do not tap) your cursor over the link. After 3 seconds, the *real* link pops up.

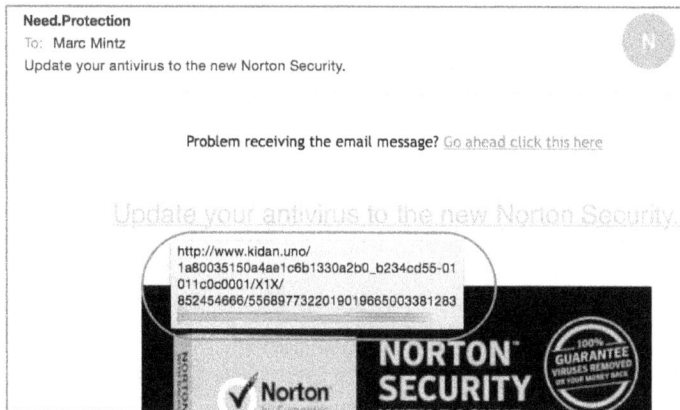

Some scams are getting more sophisticated in their choice of URL links and attempt to make them appear legitimate. For example, an email may say it is from *Bank of America*, and the link say *bankofamerica.com*, but the actual URL is *bankofamerica.tv*, or *bankofamerica.xyz.com*.

If you have any doubts at all, it is best to contact your bank, stockbroker, insurance agent, etc. directly by their known email or phone number.

13.3 Email Encryption Protocols

Three common protocols provide encryption of email between the sending or receiving computer and the SMTP (outgoing), IMAP (incoming), and POP (incoming) servers:

- **SSL**[4] (Secure Socket Layer), now depreciated
- **TLS**[5] (Transport Layer Security), the successor to SSL
- **HTTPS**[6] (Hypertext Transport Layer Secure)

Understand that these protocols only encrypt the message as it travels between your computer and your email server, and back. Unless you are communicating with only yourself (sadly, as some programmers are prone), this does little good unless you know the other end of the communication also is using encrypted email. If not, then once your encrypted mail passes from your computer to your email server, it demotes to either the less secure SSL, or if the other end of the communications does not support that, demotes to clear text from your email server, through dozens of Internet routers, to the recipient email server, and finally onto the recipient's computer. Clear text means just that: anybody who looks at that message can read it.

13.4 Transport Layer Security (TLS)

Although Secure Socket Layer (SSL) was originally considered secure, it has been broken and should no longer be used for email that is sensitive, secure, or related to healthcare, legal, government, or military sites. Transport Layer Security[7] (TLS) is the successor to SSL. To use TLS, the following criteria must be met:

- Your email provider offers a TLS. Many do not. If your provider does not offer this, *run*, don't walk, to another provider. If you are not sure which to select, I'm a fan of Google mail.

- You are using an email application as opposed to using a web browser to access your email.

- Your email application supports TLS.

---

[4] *http://en.wikipedia.org/wiki/Secure_Sockets_Layer*
[5] *http://en.wikipedia.org/wiki/Secure_Sockets_Layer*
[6] *http://en.wikipedia.org/wiki/Https*
[7] *https://en.wikipedia.org/wiki/Transport_Layer_Security*

- Your email provider has enabled and configured your email service to use TLS (they may *offer* TLS, but it may not be *enabled* by default).

- You have configured your email application to use TLS. Most email applications now do this automatically. The Apple Mail.app has gone to the point of removing the preference setting for both SSL and TLS.

- Lastly, although not a requirement for TLS, a requirement to stall off hacking your password is that your email provider should allow for strong passwords, and you have assigned a strong password to your email. As noted, I recommend 15 or more characters. Unfortunately, many providers still limit you to a maximum of 8-character passwords.

13.4.1 Assignment: Determine if Sender and Recipient Can Use TLS

Email automatically downgrades to the lowest common security protocol. In this assignment, you discover if both your own email and that of a recipient can use TLS email encryption.

In this assignment, you test if sender and recipient can use TLS

1. Open a web browser, then go to *https://checktls.com.*

2. Scroll down the home page to the *Check Your, or Any, Email System* section.

3. In the *Check How You Get Email (Receiver Test) FREE* field, enter your email address. Then select the *Check It* button.

4. The website runs tests against the domain's mail servers (MX servers), then reports on their level of security. Tap the *Show Detail* button.

MX Server	Pref	Answer	Connect	HELO	TLS	Cert	Secure	From
aspmx.l.google.com [173.194.68.27:25]	1	OK (10ms)	OK (11ms)	OK (12ms)	OK (10ms)	OK (124ms)	OK (12ms)	OK (10ms)
alt1.aspmx.l.google.com [64.233.186.27:25]	5	OK (128ms)	OK (274ms)	OK (259ms)	OK (257ms)	OK (196ms)	OK (330ms)	OK (257ms)
alt2.aspmx.l.google.com [209.85.202.27:25]	5	OK (78ms)	OK (120ms)	OK (96ms)	OK (91ms)	OK (196ms)	OK (94ms)	OK (92ms)
aspmx2.googlemail.com [64.233.186.27:25]	10	OK (123ms)	OK (281ms)	OK (255ms)	OK (252ms)	OK (202ms)	OK (313ms)	OK (255ms)
aspmx3.googlemail.com [209.85.202.27:25]	10	OK (77ms)	OK (105ms)	OK (94ms)	OK (93ms)	OK (192ms)	OK (94ms)	OK (92ms)
Average		100%	100%	100%	100%	100%	100%	100%

CheckTLS Confidence Factor for "marc@thepracticalparanoid.com": 100

5. If your *Test Results* are not 100% secure, then discuss this with your email provider for a resolution or change providers.

6. Repeat steps 1-4 using the domain of your recipient email address.

7. If the recipient's *Test Results* are not 100% secure, advise them to discuss this with their email provider or to change providers.

13.4.2 [Windows] Assignment: Install Mozilla Thunderbird

13.4.3 [Windows] Assignment: Configure Microsoft Outlook 365 for TLS

13.4.4 Assignment: Configure Mail.app for TLS

Fortunately, the Mac Mail app auto-configures for TLS or SSL if the email provider offers it. There is no manual configuration possible.

13.4.5 [Optional] Assignment: Configure Forced TLS With Paubox

One option to force bi-directional, HIPAA-compliant, secure, encrypted TLS email communications is by using Paubox[8]. Paubox is a third-party email server that can be used with your existing G-Suite, Office 365, and Microsoft Exchange email, as long as you have a private domain name associated with your email (for example, I use marc@thepracticalparanoid.com, but not marcmintz@gmail.com).

When sending email through Paubox, your interface doesn't change. You can still use the email client or web browser you always use. What does change is that your email now goes through Paubox. Paubox servers check with the recipient email server. If the recipient server supports TLS, your email is sent to the recipient as usual.

If the recipient email server does not support TLS, the email is instead stored on the Paubox server. A message is sent to the recipient inbox stating that an email is waiting for them on the Paubox server, and all they need do is tap the button to view the email.

The recipient can reply to the email, which then is sent end-to-end encrypted with TLS. But understand that if the recipient simply emails to you, and their server does not support TLS, the email is sent without end-to-end TLS encryption.

In this assignment, you create a free 14-day trial Paubox account.

- Prerequisite: An email with a private domain name, not a generic such as @gmail.com.
- Prerequisite: Access to the DNS records control panel for that private domain name.

Create a Paubox Account

1. Open a browser to *https://paubox.com.*
2. Tap the *Start for Free* button.
3. Follow the onscreen instructions to create an account and start your free trial.

Configure G-Suite

If you use Google as your email host, you must configure it to work with Paubox.

---

[8] *https://paubox.com*

4. Open a browser to *https://admin.google.com*.

5. Navigate to *Apps* > *G-Suite* > *Mail* > *Advance Settings* > *General Settings* tab > *Inbound Gateway*.

6. Select *Edit* button for Inbound Gateway. Configure as below. Then tap *Save*:

Edit setting ✕

Inbound gateway   Help

Paubox   Edit

1. Gateway IPs

IP addresses / ranges	ADD
52.26.0.0/16	

☐ Automatically detect external IP (recommended)
☐ Reject all mail not from gateway IPs
☑ Require TLS for connections from the email gateways listed above

2. Message Tagging

☐ Message is considered spam if the following header regexp matches

CANCEL   SAVE

7. Scroll down near the bottom of the page to *Outbound Gateway*. Enter *outbound.paubox.com*.

Routing

Outbound gateway   Route outgoing emails to the following SMTP server: ❓
Locally applied

outbound.paubox.com

8. At the bottom right corner of the browser, tap *Save*.

Configure DNS MX Records

9. Using a browser, log into your DNS records. In this example, it is GoDaddy.

10. Go to your *DNS records control panel*.

11. Add the following *MX* (Mail Exchanger) record:

   - Host: *@*
   - Points to: *mx2.paubox.com*
   - Priority: *10*

12. *Save* the new MX record.

13. Delete any other MX records.

14. While still in your DNS records control panel, add the following *TXT* (Text) record:

   - Host: *@*
   - TXT Value: *v=spf1 include: _spf.paubox.com include: _spf.google.com -all*

15. *Save* the new TXT record.

16. Exit the DNS records control panel.

17. If you have been following along with the Paubox onscreen instructions, tap the *Next* button. This takes you to the *Confirm your DNS record settings* page.

Send email to a TLS Recipient

18. When sending email to a recipient who supports TLS encryption, all goes as normal, except for an added footer stating *This email was seamlessly encrypted for your privacy and security by Paubox.*

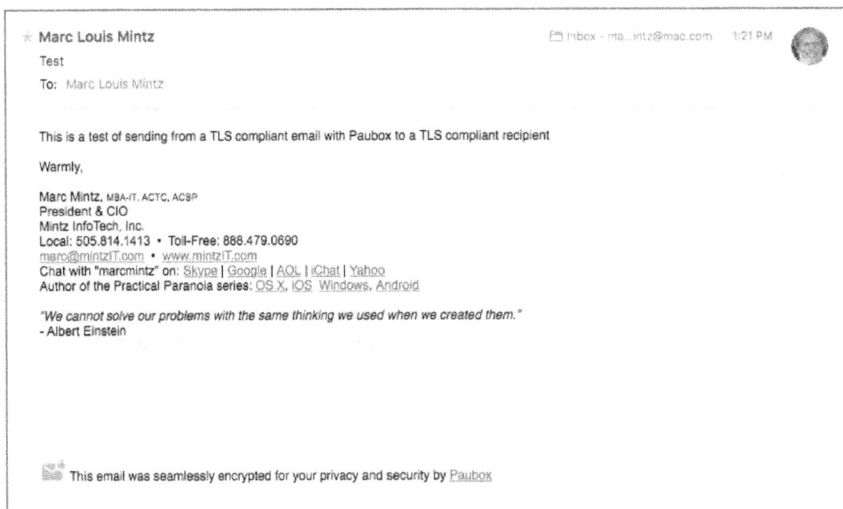

Send email to a non-TLS recipient

19. When sending email to a recipient who does not support TLS, your email is encrypted between your computer, to the Paubox mail server. Paubox then sends an email to the recipient alerting them of the mail waiting at Paubox. The recipient taps the *View Message* button. A secure browser window opens to let the recipient view the email by tapping the *View Message* button. The recipient can reply to the email from this window securely with end-to-end TLS encryption.

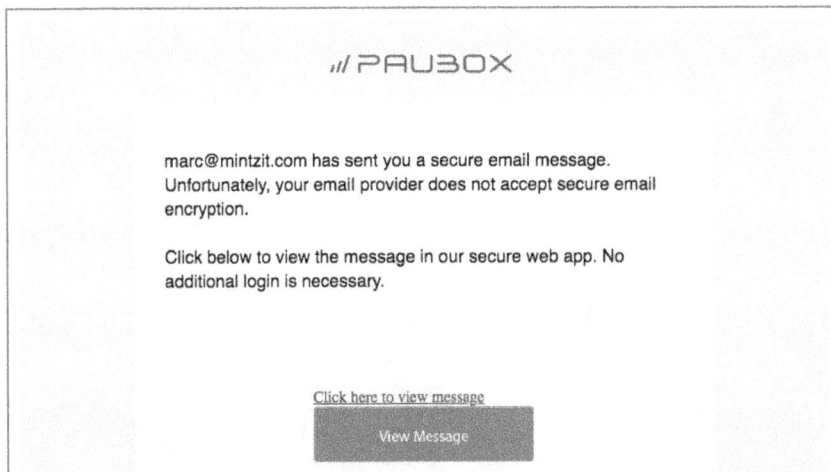

```
  Reply    (Login not required to reply)


 To: marcmintz@mailinator.com

 From: marc@mintzit.com

 Subject: Test

 Encrypted Message:                                                        🖨


      <<=======
      This is a test of sending from a TLS compliant email with Paubox to a non-TLS compliant recipient


      Warmly,

      Marc Mintz, MBA-IT, ACTC, ACSP
      President & CIO
      Mintz InfoTech, Inc.
      Local: 505.814.1413  •  Toll-Free: 888.479.0690
      marc@mintzIT.com  •  www.mintzIT.com
      Chat with "marcmintz" on: Skype | Google | AOL | iChat | Yahoo
      Author of the Practical Paranoia series: OS X, iOS  Windows, Android

      "We cannot solve our problems with the same thinking we used when we created them."
      - Albert Einstein
```

        📧 This email was seamlessly encrypted for your privacy and security by Paubox

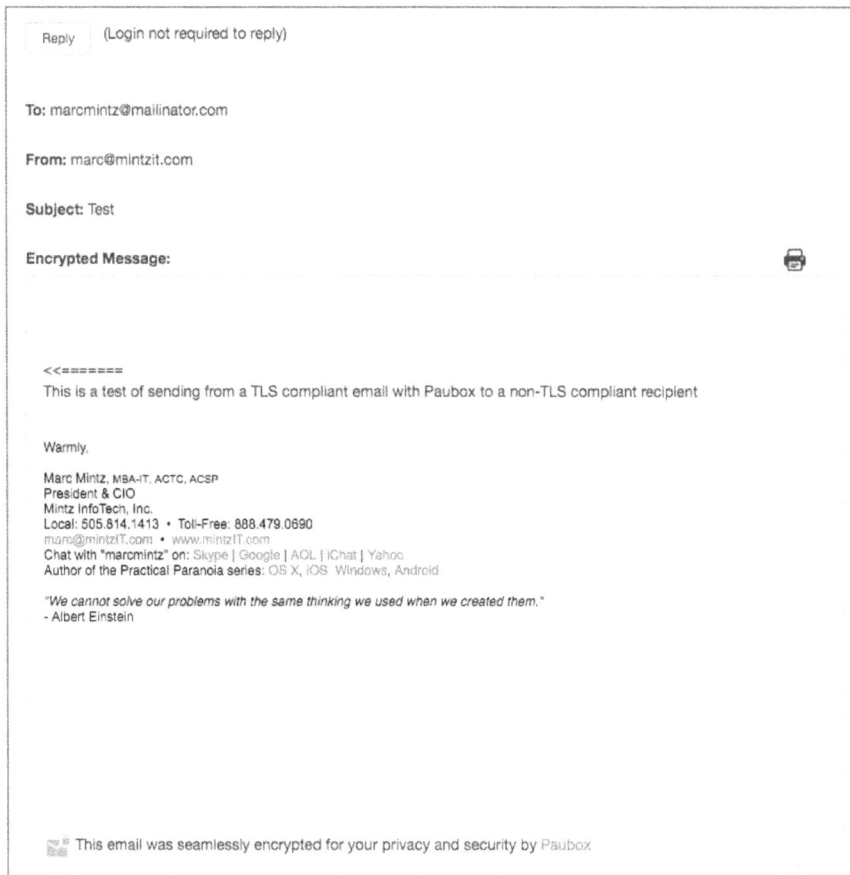

13.5 HTTPS With Web Mail

We discussed HTTPS in the previous chapter. It is an encryption protocol used with web pages. It also can be used to secure email that is accessed via a web browser. When using HTTPS, your username and password are fully encrypted between your device and the email host server, as are the contents of all email that you create or open.

When using a web browser to access email, it is vital that your email site use the HTTPS encryption protocol to help ensure data and personal security.

13.5.1 Assignment: Configure Web Mail to Use HTTPS

If using a web browser to access your email, it is critical that your web connection use HTTPS.

In this assignment, you verify that your browser-based email uses HTTPS.

- Note: If you do not use browser-based email, you may skip this assignment.

1. Launch your web browser.

2. Go to your login page for your email. In this example, we use Google Mail (Gmail).

3. As in the screen shot below, make sure that the URL field shows either the lock to the left of the URL, or *https://* and not *http://*. This lock or the https indicates you are communicating over a secure, encrypted pathway.

4. If instead, your browser shows the URL to be http://, try revisiting your email log in page, but this time manually enter `https://`.

5. If you get to the log in page, all is good. Just bookmark the https:// URL and use it instead of the previous non-secure URL.

6. If you cannot get to your log in page, change your email provider NOW!

13.6 Mail Privacy Protection

New with iOS 15 and macOS 12 is Mail Privacy Protection (MPP). With MPP active, your IP address is hidden from those sending you email, and tracking pixels are blocked. With tracking pixels blocked, email senders cannot determine your IP address if you have opened the email or gather any other information about you.

To accomplish MPP, you must opt in for the service, and use the built-in Mail.app. Once turned on, email is no longer routed directly to your computer. Instead, incoming email is routed through at least two proxy servers. The first thing the servers do is assign you a temporary IP address that is not linked to your machine or account, but is the IP address the sender sees. The IP address is associated with the geographical region you receive your email, but not your specific address.

The servers then watch for pixel trackers and then block their activity.

13.6.1 Assignment: Enable Mail Privacy Protection

In this assignment, you enable Mail Privacy Protection.

Prerequisite: macOS 12 or higher installed.

Prerequisite: Using Apple Mail.app

If Opening Mail.app For the First Time

1. Open *Mail.app*.

2. The *Mail Privacy Protection* window opens. Select *Protect Mail activity* radio button, and then tap *Continue* button:

If Mail.app Has Previously Been Opened

1. Open *Mail.app*.

2. Select *Mail* menu > *Settings* > *Privacy*.

3. Enable *Protect Mail Activity:*

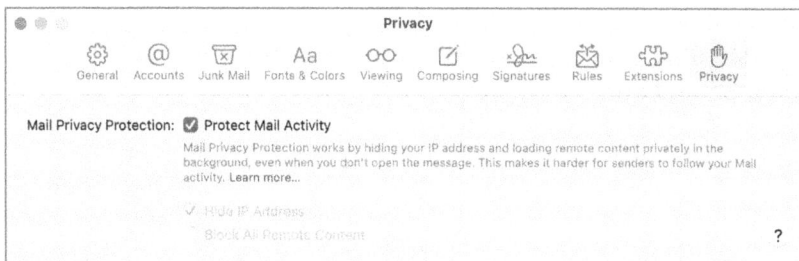

4. Close *Settings*.

13.7 Email Aliases and Hide My Email

Email Aliases

Almost all of us suffer from email overload. Although junk mail filters help, it is more band-aid than true solution. One of the fastest, easiest, and least expensive (as in *free*) solution I've found is to use a series of bogus email addresses, one for each sign up or registration you make on the web.

For example, if I'm interested in getting more product information from company xyz.com by filling out a web form, instead of using my real email address of *marc@thepracticalparanoid.com,* I may use *marc+xyz@thepracticalparanoid.com.*

Both email services that I use (Apple, with both a @mac.com and @icloud.com address, and Google, with both a @gmail.com and @thepracticalparanoid.com address) allow an unlimited number of email aliases by just adding *+sometext* between your account name and the @ sign. Your email provider may also offer this free feature.

All these aliases will be recognized as your base email account, but when receiving an email addressed to the alias, the alias will show in the email. This allows you to create rules within the mail application to manage the incoming mail. For example, should I later decide I no longer want to receive any communications from xyz.com, I can create a rule to automatically delete any incoming mail addressed to *marc+xyz@thepracticalparanoid.com.*

Below is an incoming email after subscribing for newsletters at thepracticalparanoid.com:

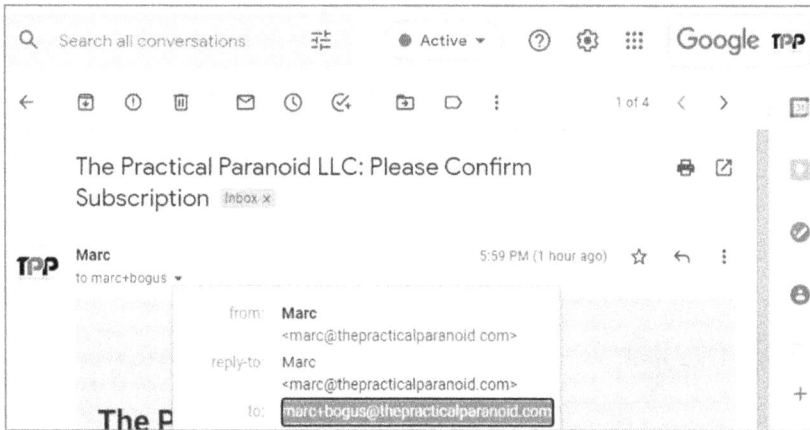

Hide My Email

New to iOS 15 and macOS 12 is *Hide My Email.* Hide My Email lets users share a unique, random email address that forwards to their personal inbox anytime they want to keep their personal email address private. This feature is built into Safari, iCloud settings, and the Mail app. It works only with Apple email accounts, such as *example@icloud.com.*

13.7.1 Assignment: Use an Email Alias

In this assignment, you determine if your email provider allows email aliases.

1. Open your email application.

2. Create a new email, addressed to an alias of your email account. For example, if your email account is *joe@icloud.com,* an alias could be *joe+test@icloud.com.*

3. Enter a subject in the subject line.

4. Send the email.

5. Wait a minute, then check for new email.

6. If your email provider allows for aliases, you will see the email in your inbox.

13.7.2 Assignment: Enable Hide My Email

In this assignment, you will enable *Hide My Email*.

● Prerequisite: macOS 12 and higher.

1. Open *System Settings > Apple ID > iCloud > Hide My Email*.

2. Select the *Options* button to the far right of *Hide My Email*.

3. At the prompt, authenticate.

4. In the *Hide My Email* window, select + (*Create New Address*):

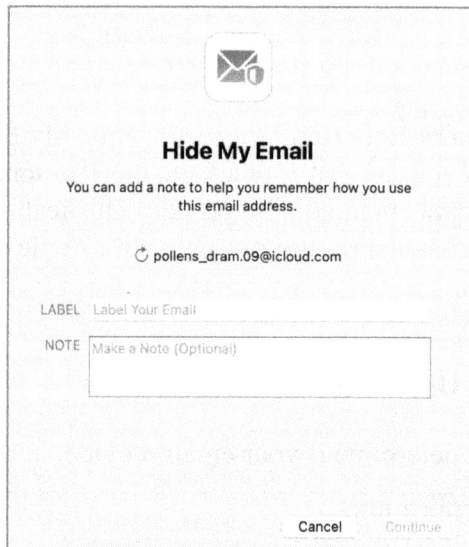

5. The *Hide My Email* window opens with a randomly generated address. If you do not like it, you can tap the cycle icon to generate others.

6. When you have an acceptable address, tap *Continue* button.

7. In the *Label Your Address* window, you can assign a *Label* and *Note* associated with this address to help remember how you use it:

8. When complete, tap the *Continue* button.

9. In the *Hide My Email* windows, you may edit your settings. When complete, select *Done* button.

13.7.3 Assignment: Use Hide My Mail

In this assignment, you will use your Find My Mail account to subscribe to a newsletter.

- Prerequisite: Completion of the previous assignment.
1. Open your browser to a site that requires an email address from you.
2. Tap inside the *Email* field. The *Hide My Email* alert will pop-up. If you would like to use *Hide My Email,* tap on the alert.
3. The *Hide My Email* window opens.
    - To use the new suggested address, tap *Continue* button.
    - To use one of your existing addresses, tap *iCloud Settings > Apple ID > Hide My Email Options,* select the *Copy* button for the desired address, then return to the web page and paste in the desired address.

13.8 End-To-End Secure Email with ProtonMail

If serious about email security, you need to use an end-to-end secure email solution. Forcing TLS for incoming and outgoing email is one option (see previous section 13.5). However, it is likely that either the sender or the recipient use email hosts which do not allow the higher security, thereby forcing TLS actions.

There are two other options for point-to-point email encryption:

- Use an email encryption utility. This works well if the other end of the communication also is using the same encryption utility. The next section covers this strategy using *GNU Privacy Guard* and *S/MIME*.

- Use a cloud-based option. This method makes it every bit as simple to send and receive email as the user is accustomed to. The downside is that instead of using an email client, a website is used to send and receive mail. An example of this is *Sendinc.com*[9].

---

[9] *https://sendinc.com/*

An interesting hybrid option is found in *ProtonMail*[10]. ProtonMail includes PGP public key/private key encryption, so that neither you nor the other party need deal with the potential headaches of installing and configuring PGP encryption.

ProtonMail offers several advantages for the typical user, including:

- Free with optional monthly/yearly plans.

- Based in Switzerland so all user data is protected by Swiss privacy laws.

- Allows the user to determine the destruction date and includes unlimited retention.

- Allows for encrypted and password protected emailing to non-ProtonMail users.

- Allows for rich text email.

When sending from ProtonMail to a non-ProtonMail user, your recipient receives an email stating that a secure message is waiting. The recipient taps the link, taking the recipient to an authentication page. Upon entering the password, the recipient then sees the message. The recipient can directly and securely reply to the message, then you receive their reply in your inbox.

When sending from ProtonMail to ProtonMail, the interface is like other email providers.

Although not as convenient as using your own email software, we find ProtonMail to be an easy choice when security, convenience, and cost are taken into consideration against the impacts of data theft, or the potential drama of confidential communications being intercepted.

13.8.1 Assignment: Create a ProtonMail Account

In this assignment, you create a free ProtonMail account.

- Note: In this assignment you create a free account. This is limited and does not allow sending encrypted mail outside of ProtonMail. You can, however, upgrade the free account at any time.

1. Open a browser to visit *https://proton.me*.

---

[10] *https://protonmail.com*

2. Select the *Create Free Account* button.

3. Select the plan you wish to use. In this tutorial, you create a free account.

4. Tap the *Get Proton for free* button.

5. Follow the on-screen instructions to create your account.

- Proton Mail always uses PGP encryption when communicating with other Proton Mail accounts. However, if you wish to use encrypted email with non-Proton Mail users, the other users will need to have PGP installed and configured on their system (no easy task), and you will need to make these changes:

6. Go to your Proton Mail page, then select *Encryption and keys* from the side bar.

7. In the main area > *External PGP settings,* enable *Sign external messages, Attach public key,* and set the *Default PGP scheme* to *PGP/MIME.*

13.8.2 Assignment: Create and Send an Encrypted ProtonMail Email

In this assignment, you send your first fully encrypted email through ProtonMail.

- Prerequisite: An existing ProtonMail account, or completion of the previous assignment

1. Open a browser to *ProtonMail* at *https://proton.me*, select the *Login* link, then log in.

2. Tap *New Message* button. The *New Message* window opens.

- Proton Mail always sends PGP encrypted messages between Proton Mail accounts. When sending to non-Proton Mail accounts that do not use PGP, your email can be encrypted with a password. When sending to non-Proton Mail accounts using PGP, your email can be encrypted with PGP and include your public key with the email

3. If sending an email to a non-Proton Mail account that does not use PGP, scroll to the bottom of the page, then tap the lock (encryption) icon. This allows you to set a password requirement to open the email from a non-ProtonMail account. Configure to your taste, then tap the *Set* button.

4. If sending an email to a non-Proton Mail account that does use PGP, scroll to the bottom of the page, tap the … button, then enable *Attach public key*.

5. Finish configuring your email, then tap the *Send* button. The program takes a moment to encrypt then send.

6. Notification of your email has been sent to the recipient.

13.8.3 Assignment: Receive and Respond to a ProtonMail Email

In this assignment, you reply to a ProtonMail secure email.

• Prerequisites: Completion of the previous two assignments.

1. If you have sent an email to a non-Proton Mail account, and have not encrypted with a password, the recipient receives an email no different from any other account.

2. If you have sent an email to a non-Proton Mail account, and have encrypted with a password, the recipient receives an email notice. To view the message, the recipient selects the *Unlock Message* button within the email.

3. After entering the required password, the email is displayed in the recipient's browser. The recipient is also able to reply via this webpage by selecting *Reply Securely*.

13.9 Outlook.com Encryption and Prevent Forwarding

Microsoft Office 365 paid subscription users with a browser based Outlook.com account have the added security and privacy features of being able to securely encrypt an email, and to prevent an email from being forwarded[11].

---

[11] *https://support.microsoft.com/en-us/office/learn-about-encrypted-messages-in-outlook-com-3521aa01-77e3-4cfd-8a13-299eb60b1957*

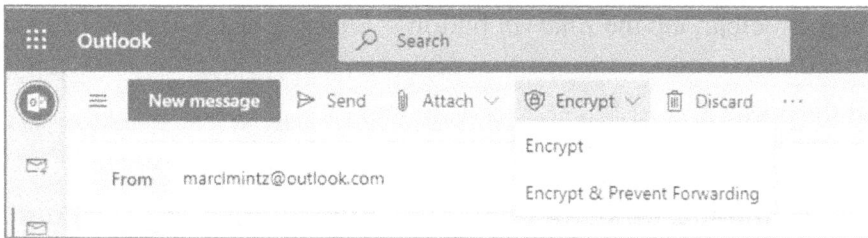

Encrypt

When creating an outgoing email with Outlook.com, the user has the option encrypt the outgoing email.

On the recipient end, any attachments may be downloaded if using Outlook.com, Outlook application for Windows 11, the Outlook mobile app, or the Mail app in Windows 11. If using a different email client, a temporary passcode can be used to download the attachments from the 365 Message Encryption portal. The email itself remains encrypted on Microsoft servers and cannot be downloaded.

Encrypt & Prevent Forwarding

As with *Encrypt* option, when selecting *Encrypt & Prevent Forwarding,* the email remains encrypted on Microsoft servers and cannot be downloaded, copied, or forwarded. MS Office file attachments (Excel, PowerPoint, Word) remain encrypted after being downloaded. If these Office files are forwarded to someone else, the other person will not be able to open the encrypted files. Non-MS Office files can be downloaded without encryption and therefore forwarded without issue.

13.9.1 Assignment: Send an Encrypted Email from Outlook.com

In this assignment, you send an encrypted email from Outlook.com.

- Prerequisite: Ownership of an Office 365 Subscription, with an Outlook.com account.

1. Open a browser to https://outlook.com, then log in with your account.

2. Create an email. Address the recipient to one of your other email addresses, or if performing this in class, to one of your study partners.

3. From the toolbar, tap the *Encrypt* button > *Encrypt*, or *Encrypt & Prevent Forwarding*.

4. Send the email.

13.9.2 Assignment: Read an Encrypted Email from Outlook.com

In this assignment, you read an encrypted email from Outlook.com.

- Prerequisite: Completion of the previous assignment.

If Using Outlook.com to Read the Email

1. Open a browser to https://outlook.com, then log in with the account set as the recipient in the previous assignment.

2. Open the encrypted email. Note that you can open, read, and reply to this encrypted email as you can with unencrypted messages.

If Using Something Other than Outlook.com to Read the Email

3. Open the email software to the account set as the recipient in the previous assignment.

4. Open the encrypted email.

5. You will see a message with instructions for how to read the encrypted message.

13.10 PGP/GPG & S/MIME

Until recently, the gold standards for email security were PGP, GNU Privacy Guard, and S/MIME. Unfortunately, configuring each is only a few steps less technically intricate than rocket science, both parties must have the same software installed, and S/MIME requires the purchase of yearly certificates by each party.

For these reasons I do not generally recommend use of these security protocols. With just 2 minutes of form entry, one can have a ProtonMail account with preconfigured PGP.

13.11   Email Validation With SPF, DKIM, And DMARC

Sender Policy Framework (SPF)

SPF[12] is an email-validation system. It provides a mechanism to authorize servers and services to send email using your domain. This allows a receiving mail exchanger (mail server) to verify that incoming mail from a specific domain is coming from a host authorized to send that mail. When a criminal hacker sends email to you with fake "from" information (for example, the sender pretends to be a vendor submitting an invoice for payment), your email server can validate or invalidate the sender as authentic. The purpose of this SPF is to prevent email with forged addresses from reaching an inbox.

If the sender of an email is validated, the email comes on through as it always has. If the sender is invalidated, this tells the receiving server to follow the rules laid out in the DMARC record (explanation below after DKIM topic) when dealing with the unauthorized email.

Domain Keys Identified Mail (DKIM)

DKIM[13] is an email authentication system to detect spoofing. It provides a mechanism for the receiver to verify that an email stating to have come from a server which has been authorized to send mail for a specific domain via SPF record, is indeed the server that is sending the email. This is done via security key exchange, the public part of which is a 2048 or larger sha256 hash algorithm encrypted key, which you will place into your DNS records in the next assignment. The DKIM purpose is to prevent acceptance of emails sent from spoofed services and servers.

DKIM works by attaching a digital signature to each outgoing email. The recipient's email system validates the signature. These signatures are normally not visible to the user.

---

[12] *https://en.wikipedia.org/wiki/Sender_Policy_Framework*
[13] *https://en.wikipedia.org/wiki/DomainKeys_Identified_Mail*

Domain-based Message Authentication, Reporting & Conformance (DMARC)

DMARC[14] is the configurable policy detailing how to deal with email that has failed the SPF and/or the DKIM validation. The easiest way to think about the DMARC process is as follows.

- SPF authorizes a server to send mail on behalf of a specific domain.
- DKIM authenticates the sending server truly is the authorized one.
- DMARC defines what to do with the email when it fails.

The options are *take no action, quarantine the email,* or *reject the email.*

13.11.1 Assignment: Configure SPF

In this assignment, you configure SPF for your email domain.

- Prerequisite: Creating SPF records is only possible if your email is on its own Fully Qualified Domain Name (such as *thepracticalparanoid.com*). If you are using a public domain such as *gmail.com, xfinity.com, myschool.edu*, you do not have the ability to edit your DNS records, and therefore cannot create your own SPF records.

- Prerequisite: Administrative access to your domain DNS records and control panel.

In this assignment, I use the domain *thepracticalparanoid.com* which is hosted with Google G-Suite, with DNS hosting at *GoDaddy.com* as the example. If your email or DNS hosts are different, the necessary steps may differ as well.

1. Open a web browser to your DNS Control Panel at your DNS host.
2. Select *Edit.*
3. Create a new *TXT* record with the following values:
   o For *Name/Host/Alias* enter @
   o For *Time to Live* enter 3600

---

[14] *https://en.wikipedia.org/wiki/dmarc*

- For *Value/Answer/Destination* enter
  `v-spf1 include:_spf.google.com ~all`

  - Note: Check with your email vendor technical support for the proper Value/Answer/Destination to replace my string.

Type *		Host *	TXT Value *
TXT	↕	@	v-spf1 include:_spf.google.com
TTL *			
1 Hour	↕		

Save   Cancel

4. Save the DNS changes.

5. Verify the DNS changes. This is done in Google from *https://toolbox.googleapps.com/apps/checkmx/*

≡ Google Admin Toolbox Check MX

Domain name

thepracticalparanoid.com| RUN CHECKS!

DKIM selector (optional)

6. Enter your domain name, then select *Run Checks!*

   - Note: Some DNS servers are not configured to send all necessary records through Google. If you are certain you have entered all your DNS records correctly and still receive an error when using the Google Admin Toolbox, try using *https://mxtoolbox.com* instead for steps 5 and 6.

7. For a detailed description of the SPF syntax, the service dmarcian.com has the most detailed, yet still user-friendly guide to understanding SPF: *https://dmarcian.com/spf-syntax-table/.*

8. If any errors are reported, discuss them with your DNS host support staff to have them resolved.

13.11.2 Assignment: Configure DKIM

- Note: Creating DKIM records is only possible if your email is on its own Fully Qualified Domain Name (such as *thepracticalparanoid.com*). If you are using a public domain such as *gmail.com*, you do not have the ability to edit your DNS records, and therefore cannot create your own DKIM records.

- Prerequisite: Possess an email account with its own Fully Qualified Domain Name (FQDN).

- Prerequisite: Administrator access to your FQDN email Control Panel.

In this assignment, I use as the example the domain *thepracticalparanoid.com* which is hosted with Google G-Suite, with DNS hosting at GoDaddy.com. If your email or DNS hosts are different, the necessary steps may differ as well.

Generate a public domain key for your domain

1. Open a browser to *admin.google.com.*

2. Select *Apps > G-Suite > Gmail > Authenticate email.*

3. Select the target domain for which you want to generate a domain key.

4. Select *Generate New Record.*

5. Select *Generate.*

6. A text box opens to display a 2048-bit key.

7. Select, then copy, this key.

Create a DKIM record

8. Open a new web page, then go to your *DNS Control Panel* at your DNS host.

9. Select *Create a new TXT* record.

10. In the *TXT Value* field, *paste* in the key created in step 6 above.

11. In the Host field, enter `google._domainkey`.

TXT		
Host *	TXT Value *	TTL *
google._domainkey	v=DKIM1; k=rsa; p=MIIBIjANBg	1 Hour

Save   Cancel

12. Save the changes made to your DNS records.

13.11.3   Assignment: Sign Email with The Domain Key

In this assignment, you configure your mail server to automatically attach the DKIM key to all outgoing email.

- Prerequisite: Completion of the previous assignment.

1. Open a browser to *admin.google.com*.

2. Select *Apps > G-Suite > Gmail > Authenticate email*.

3. Select the target domain for which you want to attach a domain key.

4. Select *Start authentication*.

13.11.4 Assignment: Configure DMARC

In this assignment, you determine what to do with incoming email that is found to be spoofed or fake.

- Prerequisite: Completion of the previous assignment.

Once DKIM is in place, a decision must be made what to do with incoming email found to be spoofed or fake. A general recommendation is this:

1. Configure DMARC to do nothing with failed validations, and to notify the administrator. When first instituting the DMARC record, use this setting to observe all authorized and unauthorized email sent over the course of a few weeks. This helps to ensure you have the correct records in place and no false positives are found.

2. Reconfigure DMARC to place failed validations in quarantine and to notify the administrator. Leave on this setting for a week or two. If false positives are

found in quarantine, research the reason and resolve. Once false positives have been eliminated, go to the next setting.

3. Reconfigure DMARC to reject failed validations.

4. Open a browser to your DNS control panel at your DNS host.

5. Create a new TXT record with the following attributes:

   o For Record Name/Host enter `is DMARC`

   o For *Value* enter on one line:
      `v-DMARC1; p=none;`
      `rua=mailto:webmaster@thepracticalparanoid.com`

      ▪ Substitute your administrator email address in place of *webmaster@thepracticalparanoid.com.*

      ▪ To send failed validations to quarantine, substitute `p=quarantine`.

      ▪ To reject failed validations, substitute `p=reject`.

TXT		
Host *	**TXT Value** *	**TTL** *
_dmarc	v=DMARC1; p=quarantine; rua=	1 Hour

Save   Cancel

6. Save your changes.

7. To test your DMARC record, go to https://dmarcian.com/dmarc-inspector/

Refer to the *DMARC Tag Registry*[15] for other available options.

13.12 2-Factor Authentication

As with all your other important online accounts, it is mandatory that each of your email accounts be secured with 2-Factor Authentication.

---

[15] *https://dmarc.org//draft-dmarc-base-00-01.html#iana_dmarc_tags*

If you have an email account that does not permit 2FA, *run,* don't walk, away from that email service.

13.13 Email Lessons Learned

☐ Phishing is the attempt to acquire your sensitive information by appearing as a trustworthy source, usually via email.

☐ Secure Socket Layer (SSL) is the original email encryption protocol, now depreciated due to vulnerabilities.

☐ Transport Layer Security (TLS) is the successor to SSL and the current email encryption standard.

☐ Hypertext Transport Layer Secure (HTTPS) is the current encryption protocol for browsing.

☐ By using checktls.com you can verify if both the sender and recipient can use TLS for secure email communications.

☐ By using Paubox you can force encrypted email communications, even if the other party does not support it.

☐ ProtonMail provides for secure encrypted email using built-in GNU Privacy Guard and S/MIME.

☐ Sender Policy Framework (SPF) is an email-validation system allowing receiving mail exchangers to verify that incoming email is coming from a host authorized to do so for that domain.

☐ Domain Keys Identified Mail (DKIM) is an email authentication system to detect spoofing.

☐ Domain-based Message, Authentication, Reporting & Conformance (DMARC) is a configurable policy detailing how to deal with email that has failed the DKIM validation.

☐ Enabling 2-Factor Authentication in your email is a vital foundation to email security and privacy. If your email provider does not allow for 2FA move to another provider.

14 Documents

No matter how paranoid or conspiracy-minded you are, what the government is actually doing is worse than you imagine.

–William Blum[1], American author, and former State Department employee

What You Will Learn in This Chapter

- Use file encryption
- View EXIF data in pictures
- Remove EXIF data in pictures
- View and edit metadata in MS Office files
- View and edit metadata in pdf files
- Redact data in pdf files

What You Will Need in This Chapter

- [Optional] Adobe Acrobat Pro
- [Optional] Microsoft Office 365
- Digital photograph in .jpg or .tiff format

---

[1] *https://en.wikiquote.org/wiki/William_Blum*

14.1   Document Security

If your documents never leave your device, and you have encrypted your storage devices using FileVault 2, there is no need to go the extra step to encrypt your documents. But should you ever need to email your sensitive data to someone else or pass sensitive data via any storage device, encrypting the data goes a long way to ensuring your peace of mind.

There are several options to document encryption, each with its own benefits and drawbacks. We discuss each here.

14.2 Password Protect a Document Within its Application

A few applications are designed with document security in mind and offer their own encryption schemes. Microsoft Office and Adobe Acrobat Pro are common examples.

Although Microsoft Office products make it an easy process to password protect your documents, prior to Office 2007 (Windows) and 2011 (Mac), it was an equally easy process to break the encryption. There are many freeware and commercial utilities that can bypass the password in older versions and open the document for reading.

Starting with Microsoft Office 2007 and 2011, Microsoft changed its encryption standard to use the secure AES-128[2] algorithm. Microsoft Office 2016 (Office 365) uses AES-256[3]. Assuming an adequate password length has been selected, some researchers estimate that it would take millions of years to brute-force crack an AES-256 password with current computing power. For security during this lifetime (famous last words), the AES-256 encryption standard should be enough to protect your documents if a strong password has been chosen.

---

[2] *https://en.wikipedia.org/wiki/Advanced_Encryption_Standard*
[3] *https://technet.microsoft.com/en-us/library/cc179125%28v=office.16%29.aspx?f=255&MSPPError=-2147217396 - About*

- Note: As of this writing, only the macOS and Windows desktop versions of any Microsoft Office software can password-protect files or open password-protected files. The online, Android, and iOS versions cannot do either of these actions.

14.2.1 Assignment: Encrypt an MS Word Document

In this assignment, you encrypt a Microsoft Word (Office 365) file. This assignment uses a Word file. The process is identical for Excel and PowerPoint files.

- Prerequisite: Microsoft Word (Office 365) installed and activated.

1. Open the target document in Microsoft Word.

2. Select *Review* tab > *Protect* > *Protect Document*.

3. The *Password Protect* dialog opens. You may set a separate password to *Open*, and to *Modify* this document. Enter a password for the desired function.

- Note: Passwords for Microsoft Office products are limited to 15 characters.

4. Re-enter the password, then tap *OK*.

5. Tap the *OK* button at the bottom right of the *Password Protect* dialog.

Your document now is protected.

14.2.2 Encrypt an Apple Pages Document

14.3 Encrypt a PDF Document

As there are only a few applications that can encrypt their own documents, chances are you will work with a file whose application cannot perform the encryption. macOS can "print" any document to pdf format, and in the process, add password-protected encryption to the pdf.

- macOS 10.12.2 and earlier print to pdf services only save the file in pdf version 1.4/Acrobat 5 format. This format uses RC4 128-bit encryption, which is considered weak, and should not be used for HIPAA, SEC, legal, or other high-security needs.

- As of macOS 10.12.3, the print to pdf service saves the file in PDF 1.6 Acrobat 7 format. This format uses AES-128 encryption, which is considered strong and may be used for HIPAA, SEC, legal, and other high-security needs.

- As of this writing with macOS 13, the print to pdf service saves the file in PDF 1.3, Acrobat 4 format. This format uses RC4 encryption, which is considered weak, and should not be used for HIPAA, SEC, legal, or other high-security needs.

- Acrobat 9 and higher use AES-256 encryption.

14.3.1 Assignment: Convert a Document to PDF for Password Protection

In this assignment, you convert a file into a PDF for encryption.

1. Open any printable document currently on your computer.

2. Select *File* menu > *Print.*

3. From the *Print* window, select the *PDF* button > *Save as PDF.*

4. In the window that opens, in the *Save As* field, name the pdf version of the document, then select the *Security Options…* button. The *PDF Security Options* window opens:

5. In the *PDF Security Options* window, enable the *Require password to open document* check box, enter a desired password in the *Password* and *Verify* fields, then select the *OK* button. Record the password(s) used.

6. Quit the current document and application.

The pdf version of the document now is encrypted. If the original document no longer is needed, it may be trashed.

14.3.2 [Windows] Assignment: Convert an MS Word Document to PDF for Password Protection

14.3.3 [iOS] Assignment: Encrypt a PDF

14.4 Encrypt a Folder for macOS Use Only

An easy way to securely send a file or folder is to use a utility to archive (compress to a single file) the files or folder, and have that same utility protect the archive with a password.

macOS has a built-in utility to do this for you: *Disk Utility*. The only downside is that the archives created with Disk Utility only are readable on another macOS computer. The archives are not cross-platform compatible, but if your documents are passed along only to others using macOS, it is an excellent tool.

14.4.1 Assignment: Create an Encrypted Disk image

In this assignment you create an encrypted disk image to store sensitive files and folders.

1. Open Disk Utility, located in */Applications/Utilities*.

2. Select Disk Utility *File* menu > *New Image* > *Blank Image…*

3. Configure the *New Image* screen as below.

- Note: When assigning *Encryption,* you are prompted to create a password. For this assignment, use *password.*

- *Save As:* Enter the name for the archive that will be password protected.

- *Where:* Navigate to where you want the archive to be saved.

- *Name:* Enter the name of the mounted disk image. To avoid confusion, the mounted disk image name is normally the same as the *Save As* field. For demonstration purposes, we name them differently in this example.

- *Size:* This must be larger than the total size of files the archive will hold.

- *Format:* Mac OS Extended (Journaled). This is the macOS standard format.

- *Encryption:* 256-bit takes more time to encrypt and decrypt than 128-bit but is also more secure. When selecting this option, you are prompted to provide a password. Enter your desired password, then tap *OK*.

- *Partitions:* Single Partition, GUID Partition Map. This is the macOS standard.

- *Image Format: Sparse Bundle Disk Image.* This is the format that compresses out all unused space.

4. Select the *Save* button.

5. The archive is saved. Disk image (the opened format of the archive) is displayed in the Finder window sidebar *and* depending on your *Finder Preferences* menu > *General* > *Hard Disks,* may display as mounted on the Desktop. You now have an encrypted, password protected archive, but it is empty. Time to fill it.

6. Locate the mounted disk image on the Desktop. In our example, the disk image is called *Top Secret Files*.

7. Drag the various files and folder that you have targeted for password protection into the mounted image.

8. Eject/unmount the mounted image. It closes, removing itself from the Desktop, leaving just the password-protected archive in the location you specified in step 3 above (Desktop).

This archive may be securely passed to macOS users by any method. If a user knows the password, double tapping the archive mounts the disk image to the user's Desktop, allowing full read and write access to the documents inside.

14.4.2 [Windows] Assignment: Create an Encrypted Windows Folder

14.4.3 [Windows] Assignment: Backup Encryption Certificate for the Encrypted Folder

14.4.4 [Windows] Assignment: Import Previously Saved File Encryption Certificates

14.5 Encrypt a File or Folder for Cross Platform Use with Zip

If you need to exchange a file or files with others regardless of the computing platform used, we need to encrypt our archive in a format that is readable by any OS. There are over a dozen cross-platform encryption formats, *zip* has become the most common standard and is built into both macOS and Windows. By default, the zip encryption standard is not considered strong unless using AES-128 or AES-256 encryption. Our favorite tool for this is *7-Zip*.

Although macOS has the built-in ability to create zip archives, it uses the default zip format which lacks encryption. To encrypt zip archives, you need a third-party utility. We recommend using *Keka* for macOS.

Once you have created an encrypted archive of your file or files, the archive can be uploaded to a file server, shared by email, or passed along via drive, disc, or thumb drive.

- Note: The encryption protocol used in *zip* is considered weak and should not be used for HIPAA, SEC, legal, or other high-security needs unless using a third-party zip utility that provides AES 256-bit encryption. *WinZIP* is the industry leader for cross-platform commercial software that provides this level of *zip* security. *7-zip* is the industry leader for open-source software with this level of zip security. *Keka* uses *7-zip* as well as *zip with AES 256.*

14.5.1 Assignment: Encrypt a File or Folder Using Zip

In this assignment, you encrypt a file using Keka. The same process can be used to encrypt a folder full of items.

- Note: Keka is available for $3.99 from the Apple App Store, or downloaded for free from the developer website *https://www.keka.io.*

Download and install Keka

1. To download from the Apple App Store, open the *App Store app,* search for *Keka,* then download, or download for free from *https://www.keka.io.*

Configure Keka Preferences

2. Select the *Keka* menu > *Settings*.

3. Select the *General* tab and configure to your taste. Shown below are my configuration preferences.

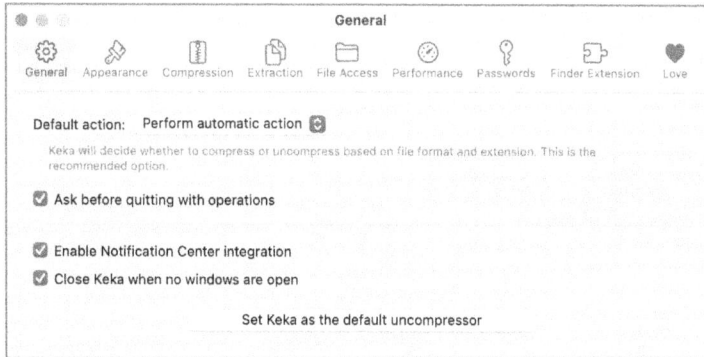

4. Select the *Set Keka as default uncompressor* button.

5. At the *An extension is required* alert, select the *More Information* button.

6. A browser opens to the Keka developer site. Tap the *Download KekaDefaultApp* button.

7. When the app completes downloading, drag it from the *Downloads* folder to the *Desktop*.

8. Double tap the *KekaExternalHelper.app* to open it. The app installs. To complete installation, tap the *OK* button.

9. In Keka preferences, tap *Appearance* tab and configure to your taste. Shown below are my settings:

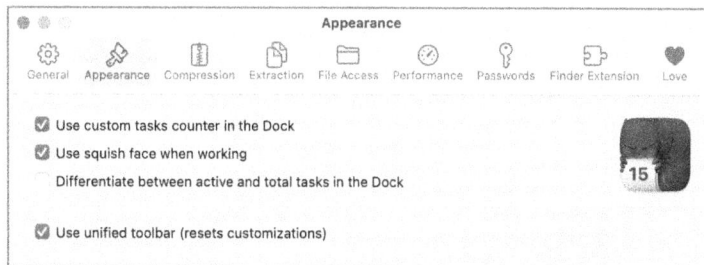

10. In the Keka preferences, tap *Compression* tab and configure to your taste. Shown below are my settings:

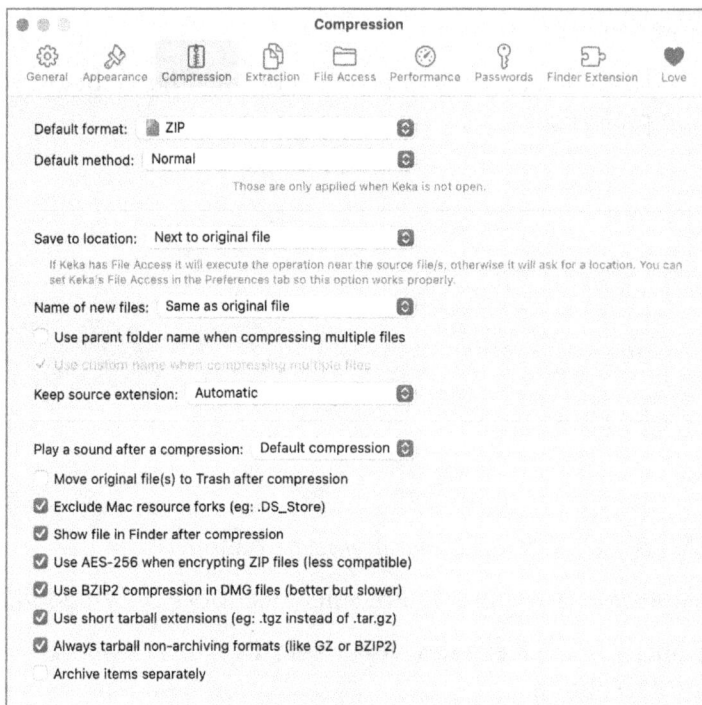

11. Select the *Extraction* tab and configure to your taste. Shown below are my preferences.

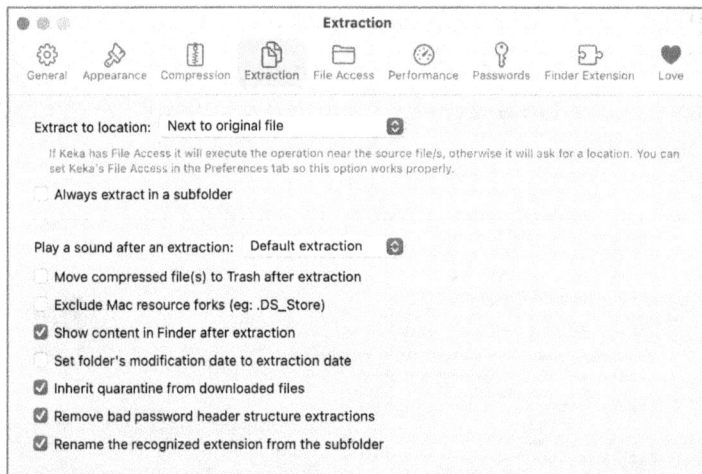

12. Select the *File Access* tab and configure to your taste. Shown below are my preferences.

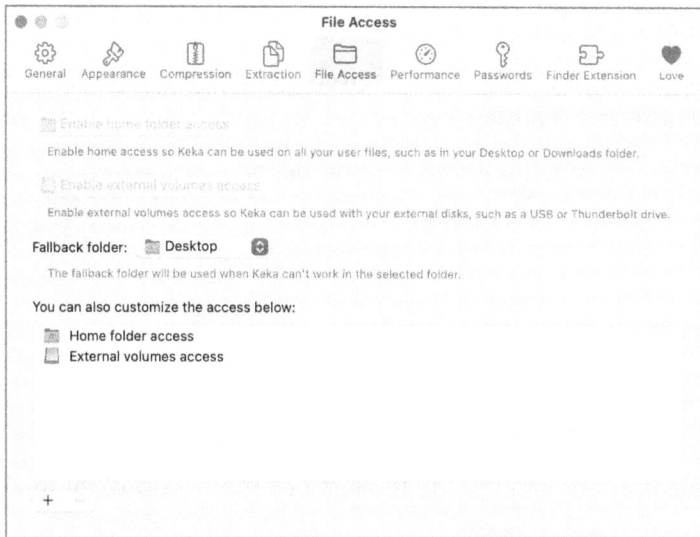

13. Select the *Performance* tab and configure to your taste. Shown below are my preferences.

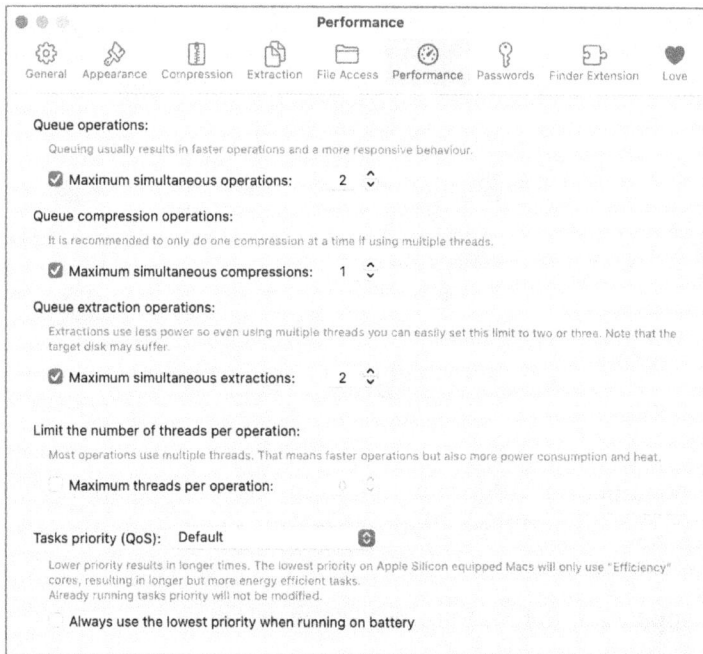

14. Select the *Passwords* tab and configure to your taste. My preference is to not use a default password.

15. Select the *Finder Extension* tab.

16. Tap *The Extension is disabled* button to begin enabling the *Keka* extension. Follow the onscreen instructions.

17. Configure the rest of the window to your taste.

18. Close *Keka Preferences* window.

19. Quit *Keka*.

Compress and encrypt a file

20. Open *Keka*.

21. In the *Keka* main window top right dropdown menu:

    o   From the pop-up menu in the top right of the window, set to *Zip*.

    o   Set Method to Normal

    o   Enter the desired password in the *Password* and *Repeat* fields

    o   Enable *Use AES-256 encryption* checkbox.

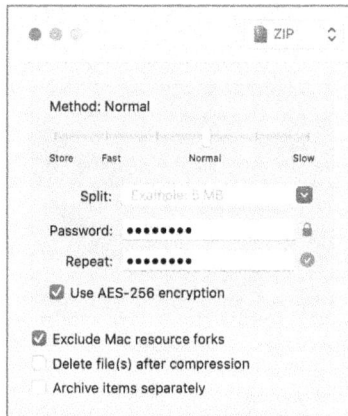

22. Locate the document or folder on your computer to be compressed and encrypted, then drag and drop it onto the *Keka Dock* icon. For my example, the file is named test.png.

23. A .zip file appears in the same folder as the original.

14.5.2 Assignment: Set Keka as the Default for Zip

macOS cannot recognize AES 256 encrypted .zip files, but *Keka* can! To open these files, you must train macOS to use *Keka* instead of the built-in default zip utility.

- Prerequisite: Completion of the previous assignment.

1. Tap to select the encrypted zip file created above.

2. Select the *File* menu > *Get Info*.

3. If not currently visible, expand to view the *Open with* area. Note that the default *Archive Utility.app* is selected. Select the *Open with* pop-up menu, select *Keka.app*, then select *Change All...*

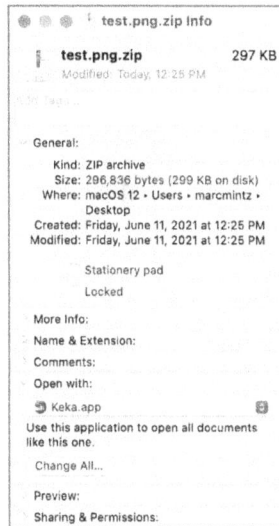

4. Close the *Get Info* window.

5. From now on all .zip files open with *Keka* (including those encrypted that macOS cannot open natively).

14.5.3 Assignment: Open an Encrypted Zip Archive

In this assignment, you open the encrypted zip archive created in the previous assignment.

- Prerequisite: Completion of the previous assignment.

1. Locate the *encrypted zip archive* created in the previous assignment.

2. Tap the *Extract* button.

3. Double tap on the *encrypted zip archive*.

4. At the prompt, enter the password used to encrypt the archive, and then tap *Done* button.

5. The archive opens, saving the contents to the same folder as the zip file.

14.6   Exchangeable Image File Format and Metadata

The Exchangeable image file format[4] (Exif) is a standard for storing metadata for image (JPEG, PNG, HEIC, TIFF, WAV, but not JPEG 2000 or GIF) and sound files. *Metadata*[5] is "information about the information." Typically, the metadata is stored within the file itself, but the standard does allow for the Exif data to be stored separately from the file.

The following image data may be stored:

- Camera settings: aperture, focal length, ISO speed, metering mode, and shutter speed

- Copyright information

- Creation date and time

- GPS (Global Positioning System) coordinates

- Image thumbnail

- File description.

Developers may opt to include additional image information.

---

[4] *https://en.wikipedia.org/wiki/Exif*
[5] *https://en.wikipedia.org/wiki/Metadata*

The following audio data may be stored:

- Bits per sample
- Bytes per second (average)
- Date createdEncoding format
- Exif version
- Make
- MakerNote
- Model
- Number of channels
- Related image file
- Sampling rate
- Time created

Developers may opt to include additional audio information.

The issue with Exif is the embedded GPS data. This value can place the exact location on earth where the photograph was taken. Add the creation date and time and there exists a 4-dimensional positioning for where and when the photograph was taken.

As an example of how this affects security and privacy, let's say you take a picture of your fine sushi dinner and post it. As most websites and social media do not strip out Exif data, anyone viewing the image can know that, based on the date/time stamp and GPS, you took the photo 5 minutes ago at a restaurant 100 miles from your home. A couple of hours allow more than enough time to rob your home and get away.

14.6.1 Assignment: View Exif Data in a Photograph

In this assignment, you view the Exif data in a photograph.

- Prerequisite: A digital photograph in JPG, PNG, or TIFF format on your computer. If you do not have a file in this format, take a picture with your phone or tablet then send it to yourself.

Use Get Info to view Exif

1. Right-tap > *Get Info* submenu on a digital photograph on your computer. The *Get Info* window appears.

2. Inside the *Get Info* window, in the *More Info* area, you can view most Exif data.

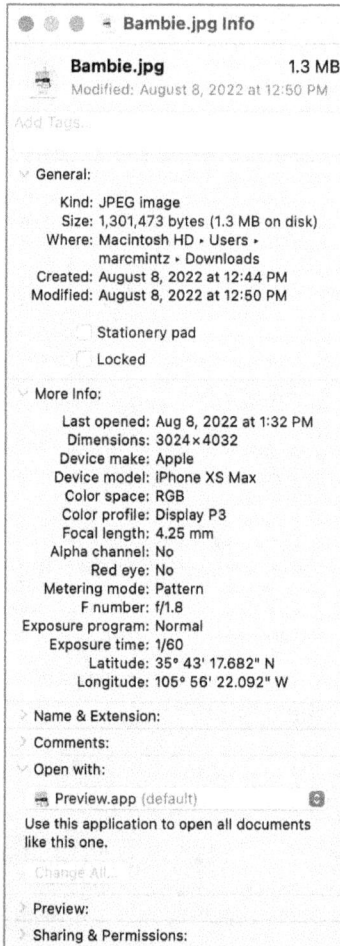

3. Close the *Get Info* window.

Use Gallery mode to view Exif

The *Finder Galley* and *Finder Column* modes often display more of the Exif data.

4. Within the *Finder*, navigate to the folder containing the target digital photograph.

5. Tap the folder icon.

6. In the Toolbar, tap on the *View > as Gallery* or *Column* icon.

7. Select the target digital photograph.

8. Near the top right quadrant of the window, to the right of *Information,* tap *Show More.* All available Exif data displays in the right sidebar.

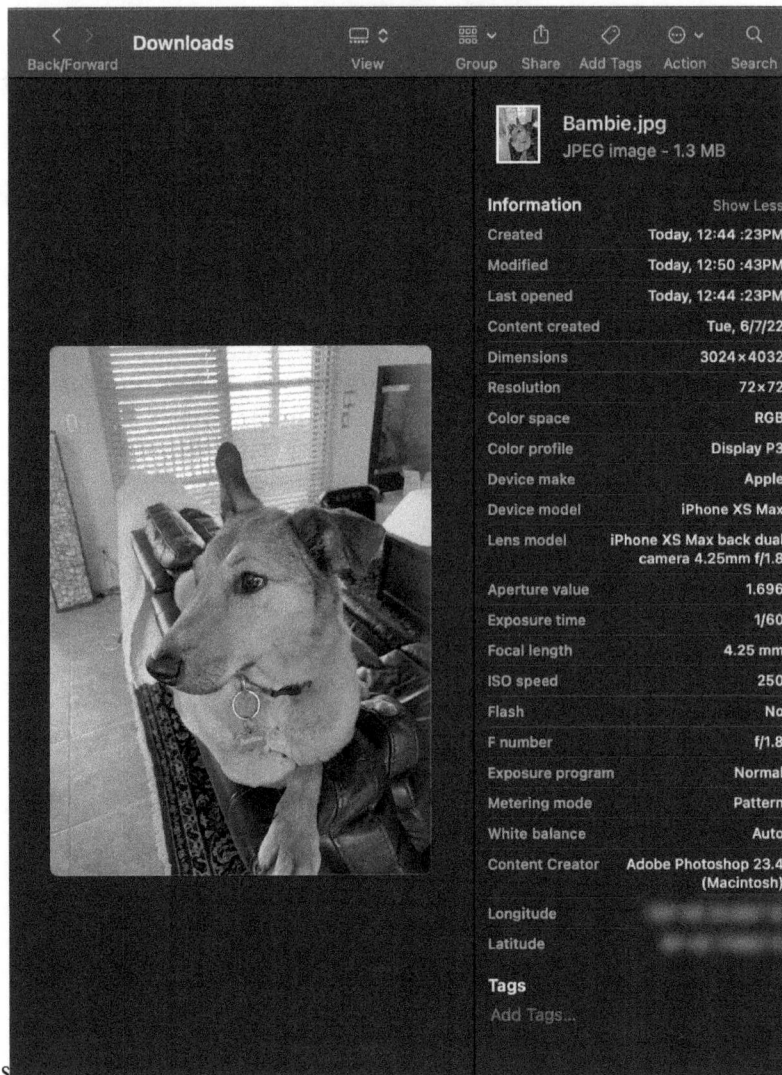

9. Close the *Finder* window.

Use Preview to View Exif Data

macOS includes a built-in graphics viewer named *Preview*. It also allows access to viewing Exif data.

10. Open the photograph into *Preview*.

11. In Preview, select *Tools* menu > *Show Inspector*.

12. In the Inspector window, select the *Info* tab. You now have four tabs of Exif data–*General, Exif, GPS,* and *TIFF.* Explore the information provided in each:

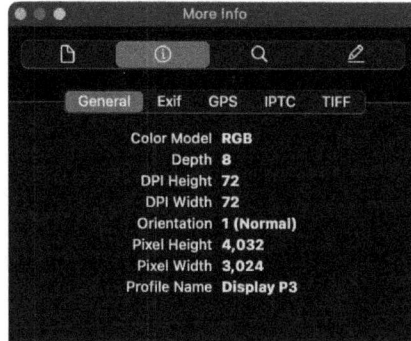

14.6.2 Assignment: Edit and Delete Exif Data

GPS location is one of many data points that can be isolated from Exif and affect your privacy. Before you distribute images, you can remove some or all your Exif data.

Fortunately, macOS includes a tool to do this–*Preview.*

In this assignment, you edit or delete Exif data from your image in Preview.

● Prerequisite: A digital photograph.

1. Open the digital photograph into Preview.

2. Select *Tools* menu > *Show Inspector.*

3. In the *Show Inspector* window, select the *More Info* tab > *GPS* tab:

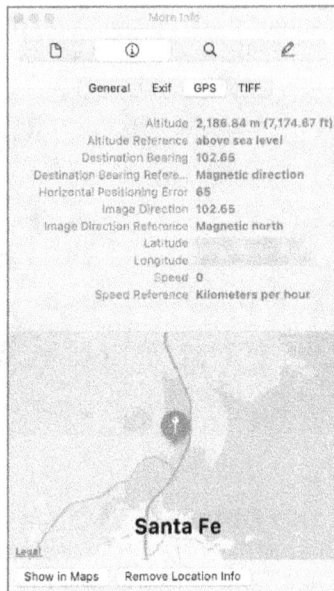

4. Tap *Remove Location Info* button.

5. Note how all GPS location information has been removed from the file.

14.6.3 Assignment: View and Edit Metadata from an Office 365 Word File

Microsoft Office is one of the most common file formats for exchanging information. All files hold metadata that may be considered sensitive or private. In the case of legal documents and medical records, users may encounter lawsuits and fines for including metadata in these files.

In this assignment, you remove all metadata from a Microsoft Office 365 *Word* file.

- Prerequisite: Access to a Microsoft Office 365 *Word* file.

1. Create a copy of your file. You will work on the copy so as not to modify the original.

2. Open the copy into MS Word.

3. Select *File > Properties* tab.

4. Delete any information you do not want to appear in the *Word* file, then tap *OK*.

5. Tap the *Review* tab.

6. Turn off *Track Changes*.

7. To the right of *Track Changes,* tap the pop-up menu to select *All Markup*.

8. Under *All Markup,* tap *Markup Option,* then enable *Comments, Insertions and Deletions,* and *Formatting*.

9. If the file has any tracked changes (displayed in the right sidebar):

   - Review each tracked change by either selecting *Review* tab > *Accept* icon > *Accept and move to next* or by selecting *Review* tab > *Reject* icon > *Reject and move to next*. Repeat until all tracked changes are cleared.

   - Accept all tracked changes by selecting *Review* tab > *Accept* icon > *Accept all changes*.

   - Reject all tracked changes by selecting *Review* tab > *Reject* icon > *Reject all changes*.

10. If the file has any comments (displayed in the right sidebar), either:

    - Delete all comments by selecting the *Review* tab > *Delete* (under *New Comment*) > *Delete all comments in document*. Or,

    - Review each comment before deletion by selecting the *Review* tab > *Next* (to the right of *New Comment*) > then *Delete*.

11. When complete, *Save* the file.

14.6.4 Assignment: Remove Metadata from an Office 365 Excel or PowerPoint File

- Prerequisite: Access to a Microsoft Office 365 *Excel* or *PowerPoint file*.

1. Create a copy of your file. You will work on the copy so as not to modify the original.

2. Open the copy.

3. Select *File* > *Properties* tab.

4. Delete any information you do not want to appear in the *Excel* or *PowerPoint* file, then tap *OK*.

5. When complete, *Save* the file.

14.6.5 Assignment: Redact Content in a PDF File

If you have sensitive data within a pdf file, Adobe Acrobat Pro (for-fee software) provides a tool to redact (black out) and remove this data.

- Prerequisites: Installation of Adobe Acrobat Pro and a pdf file.
- Note: Adobe offers a free trial of Acrobat Pro. If you prefer you can also install a free trial version of *PDF Editor 12* from Foxit.com to use with this assignment.

1. Open your pdf in *Adobe Acrobat Pro*.

2. Tap the *Tools* tab > *Redact*.

3. Highlight the text or image to be redacted.

4. Tap the *Apply* button.

5. Repeat for all items to be redacted.

14.6.6 Assignment: View Object Metadata in a PDF File

Metadata in a PDF takes three forms:

- *Properties*. This includes file name, author name, subject, and keywords.
- *Object metadata*. This is metadata embedded in a file that is embedded in the pdf, typically a photograph.
- *File revisions*. This includes changes to the text of the file.

If an object saved in a pdf has metadata attached to it, Adobe Acrobat Pro allows you to see the metadata.

- Prerequisites: Installation of Adobe Acrobat Pro, and a pdf file containing an object with metadata.

- Note: Optionally, you may install a trial version of *PDF Editor 12* from Foxit.com to use with this assignment.

1. Open your pdf in *Adobe Acrobat Pro*.

2. From *Tools,* select *Edit PDF > Edit Text & Images*.

3. Tap on an object, then right-tap the selection > *Show Metadata*. If metadata is available, it displays. If no metadata is available, the *Show Metadata* submenu does not display.

14.6.7 Assignment: View and Edit Document Metadata in PDF

In this assignment, you read the document metadata of a PDF file and change metadata.

- Prerequisite: Adobe Acrobat Reader or Adobe Acrobat Pro installed.

- Note: Optionally, you may use the free web service *2pdf.com* to edit the metadata.

1. Open a target PDF file into either Adobe Acrobat Reader or Adobe Acrobat Pro.

2. Tap *File* menu > *Properties*. The *Properties* window opens, displaying the document metadata.

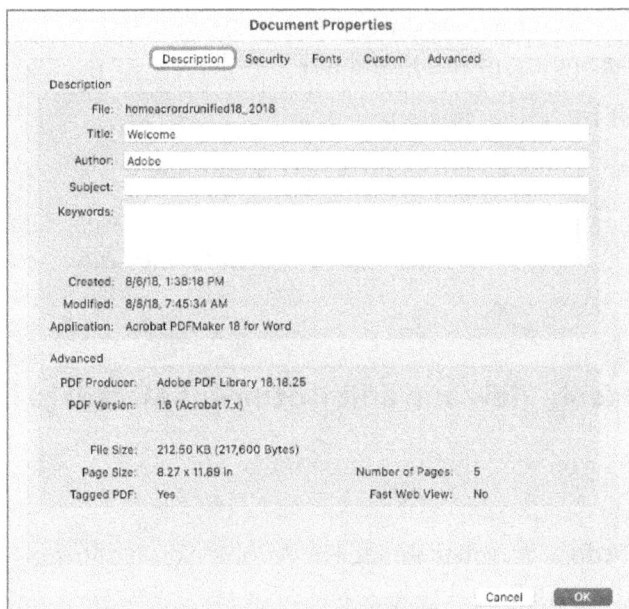

3. If you do not have *Adobe Acrobat Pro* installed, skip to step 8. If you have *Adobe Acrobat Pro* (Adobe Reader only allows viewing metadata), select the *Author* field, then enter your name.

4. Repeat for other fields.

5. Tap *OK.*

6. Tap *File* menu > *Save As,* then save your new file with modified metadata.

7. Quit Adobe Acrobat Pro, and then continue with the next assignment.

8. If you do not have *Adobe Acrobat Pro*, open a browser to *https://2pdf.com.*

9. Tap the *Edit PDF metadata* icon.

10. Tap the *Choose file* button.

11. Navigate to locate your target pdf.

12. The 2pdf.com page displays the current document metadata.

13. Edit the metadata to your taste.

14. Tap the *Update PDF Metadata* button.

15. Tap the *Download* button. The pdf document downloads to your iOS device with the edited or deleted metadata.

16. As you are not able to view metadata on iOS, tap the *Send* icon at the bottom of the screen, then email the pdf to yourself.

17. Open the emailed pdf on a computer with Adobe Acrobat Reader.

18. In Acrobat Reader, select *File* menu > *Properties* > *Description.* The document metadata will display.

19. Note how the metadata now displays with your modified data.

20. Quit Adobe Reader.

14.6.8 Assignment: Remove All Metadata in a PDF File

In this assignment, you remove all metadata from a pdf.

● Prerequisite: Adobe Acrobat Pro installed.

● Note: Optionally, you may install a trial version of *PDF Editor 12* from Foxit.com to use with this assignment.

● Prerequisite: PDF file.

1. Open *Adobe Acrobat Pro.*

2. Open the target file.

3. Perform the previous assignment on this file.

4. Select *File* menu > *Save As Other* > *Optimized PDF.*

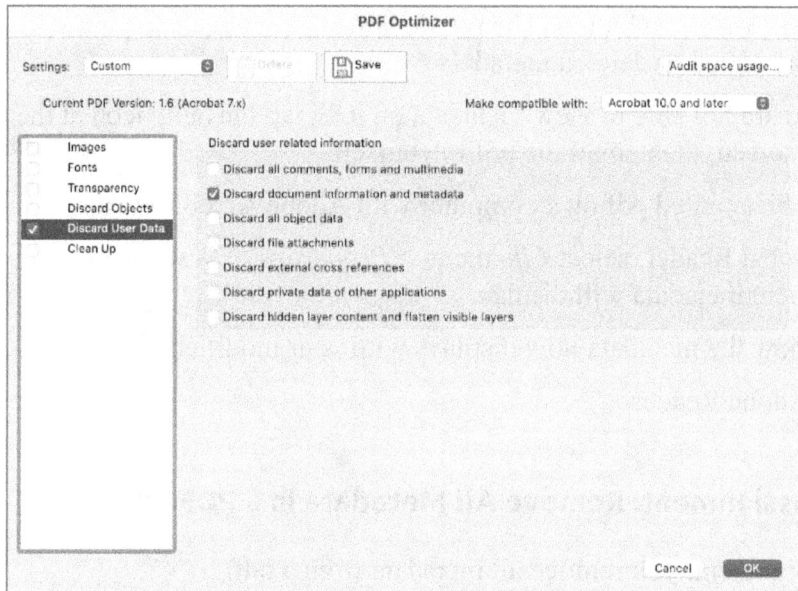

5. Enable *Discard User Data,* disable all other sidebar options.

6. Select *Discard User Data.*

7. *Enable* Discard document information and metadata. Disable or enable other elements to your taste. I generally recommend enabling (discarding) all.

8. Tap *Save.* Name the new optimized file, then save to desired location.

14.7   [Android] Secure Folder

14.8   Documents Lessons Learned

☐ Microsoft Office 365 macOS and Windows desktop versions can encrypt its files using AES-256. This is the current industry standard. Earlier versions used vulnerable encryption.

☐ macOS 12 *print to pdf* services saves a file in PDF 1.6 Acrobat 7 format, which uses AES-128 encryption. Although this is not as strong as AES-256, it is accepted for HIPAA, SEC, legal, and other high security needs.

☐ Although macOS 13 has built-in ability to compress files and folders using *zip*, it does not have the ability to encrypt these zip archives.

☐ *Keka* is a macOS-only freeware and shareware that can encrypt and decrypt using the current industry standards.

☐ Exchangeable image file format (Exif) is a standard for storing metadata for jpeg, tiff, wav image files and sound files.

☐ The macOS Get Info window displays some of the Exif data but does not provide edit capabilities.

☐ The Preview app can be used to remove GPS metadata from a digital image.

☐ To remove metadata from an Office 365 Word file: 1) Open File menu > Properties tab, then delete unwanted information. 2) Tap the Review tab > turn off Track Changes > tap All Markup > tap Markup Option > enable Comments, Insertions and Deletions, and Formatting. 3) If the file shows any tracked changes in the sidebar, review then accept or reject changes. 4) If the file has comments displayed in the sidebar, delete the comments. 5) When complete, save the file.

☐ PDF files may contain three types of metadata: properties, object metadata, and file revisions.

15 Voice, Video, and Instant Message Communications

Surveillance technologies now available–including the monitoring of virtually all digital information–have advanced to the point where much of the essential apparatus of a police state is already in place.

- Al Gore[1]

What You Will Learn in This Chapter

- Install and configure *Signal*

- Secure instant message with *Signal*

- Secure voice and video calls with *Signal*

What You Will Need in This Chapter

- No additional resources required.

15.1   Voice, Video, and Instant Messaging

Every time you send or receive a text message, phone call, or videoconference on your computer or mobile device, your conversations and metadata are stored by third parties. The manufacturers or developers (such as Apple, Facebook, Google, etc.) and carriers (Verizon, AT&T, etc.) for each party can intercept any traffic that crosses their networks. This interception may extend to any third parties that

---

[1] *https://en.wikipedia.org/wiki/Al_Gore*

work with your carrier, such as contractors, or subsidiaries. In addition, your local, state, and federal government monitor data in dragnet-style snooping.

How can you communicate easily and securely?

If you are interested in cross-platform, end-to-end encrypted, text, voice and video conferencing solutions, a few options are available.

Introduced with macOS 12 and iOS 15 is Apple's FaceTime messaging service on Android and web, allowing Windows and other users share in secure messaging with the rest of us. Although Facetime is encrypted, Apple apparently does have some limited back door access. Because of this, we can only give FaceTime a 9.5 out of 10 in the cybersecurity rating.

Wire[2] and *Signal*[3] are our choices for end-to-end encrypted voice, video, instant messaging, and group communications. Both provide end-to-end encrypted communications between Android, Chrome OS, iOS, macOS, and Windows.

Wire is a for-fee commercial service. It offers a free 30-day trial.

Signal is an independent nonprofit that provides its product and services for free. We use *Signal* for the rest of this chapter.

15.2   HIPAA Considerations

HIPAA is concerned about securing *Protected Health Information* (PHI) from leakage, but at the same time, requires that instant messaging have an audit trail. This requires that all messaging be logged to a centralized server so the log can be reviewed. In addition, HIPAA requires that the vendor be willing to sign a *Business Associate Agreement*[4] (BAA). As a BAA puts the vendor at a potential liability should their service or software be found responsible for leaking protected health information, you will not find free or inexpensive software that meets HIPAA compliance requirements.

---

[2] *https://wire.com/*
[3] *https://www.signal.org*
[4] *http://www.hhs.gov/hipaa/for-professionals/covered-entities/sample-business-associate-agreement-provisions/index.html*

Most readers and students of this work want to leave *no* record of an encrypted conversation. Also, most of our readers have no need of a BAA.

If your instant messaging needs include HIPAA compliance (this requires meeting Joint Commission guidelines), then the rest of this chapter does not apply to you. I recommend you perform an internet search to find and assess the few options available. Then work with an IT expert to implement your HIPAA-compliant program.

15.3 Signal

Signal is a free platform for peer-to-peer (no centralization) and group secure, end-to-end encrypted communications using instant messaging, voice, and video.

15.3.1 Assignment: Install Signal

In this assignment, you create a *Signal* account. This account allows you to make fully secure, encrypted instant messaging, voice calls, and video conferences with friends and business associates.

- Prerequisite: An existing Signal account registered on an Android or iOS mobile device (performed in this assignment).

Download and install Signal onto a mobile device

Prior to installing *Signal* on a computer, the user must have *Signal* installed on an Android or iOS mobile device, then create a *Signal* account.

1. On your iOS or Android mobile device, open a browser window to *https://signal.org*.

2. Tap *Get Signal*. If using an iOS device, the *App Store* opens to *Signal-Private Messenger*. If using an Android device, the *Google Play Store* opens to *Signal-Private Messenger*.

3. Download and Install *Signal* to your mobile device.

4. On your mobile device, open the *Signal* app.

5. Follow the onscreen instructions to complete the registration process.

Download and install Signal onto a Macintosh

6. Open a browser and go to *https://signal.org,* then tap the *Get Signal* button.

   • Note: Do not download *Signal* from the App Store. None of these are the real *Signal*.

7. Open the downloaded .dmg file, it mounts a *Signal* volume on the desktop.

8. Inside the Signal volume, drag the *Signal*.app into your Applications folder.

9. Launch *Signal*.

10. Signal displays a QR code.

11. If using an iOS mobile device, open *Signal.app > Signal Settings > Linked devices > Link New Device*. If using an Android mobile device, tap the + button.

12. Use your mobile device to scan the QR code.

13. Assign a name for your Linked Device, then tap *Finish*.

Your *Signal* desktop app is now ready to use!

15.3.2 Assignment: Invite People to Signal

Before you can communicate with someone else using *Signal* they must also have a *Signal* account.

In this assignment, you invite someone to install *Signal* and create an account.

• Prerequisite: Completion of the previous assignment, or a completed installation of Signal on both a mobile device and your computer.

• Prerequisite: Access to your mobile device with Signal installed.

1. Open *Signal* on your phone (invitations do not yet work with *Signal Desktop*.)

2. Tap your *profile picture* in the top left corner > *Invite Your Friends*.

3. Select to send either a *Message* or *Mail*.

4. A list of all your phone contacts appears. Select the target contact(s), then tap *Done*.

5. A new email message is created with each of your target contacts listed in the Bcc field, with a link to download *Signal* on their phone.

6. Customize the email to your taste, then tap the *Send* button.

7. Once your target contacts have installed *Signal* on their phone, you receive a text from *Signal* they have joined, and their name appears in your *Signal Contacts* list.

15.3.3 Assignment: Secure Instant Message with Signal

In this assignment, you instant message your new *Signal* friend.

1. Open *Signal* (for this assignment, on your computer.)

2. From the sidebar, select the desired *Contact*.

3. In the main body area of the *Signal* window, at the bottom in the *Send A Message*, enter a text message for your contact, then tap the *Return* key. The message is sent to your contact and received in seconds.

15.3.4 Assignment: Secure Voice or Video Call with Signal

In this assignment, you make a secure, encrypted voice call to a *Signal* friend.

1. Open *Signal*.

2. Select a *Signal* contact to call.

3. In the top right corner of the *Signal* window tap either the *phone* or the *video* icon.

4. Tap the *Start Call* button.

5. On your friends *Signal* device, they hear their device ringing, and an *Incoming Call* message in *Signal*. If they wish to answer, they tap the *Signal Phone* icon.

6. The two of you can now speak in complete privacy (even better than Maxwell Smart's Cone of Silence[5]).

---

[5] *https://en.wikipedia.org/wiki/Cone_of_Silence*

7. To disconnect, either party taps the *Phone* or *Video* icon.

15.4 Voice, Video, and Instant Message Communications Lessons Learned

☐ Phone calls, videoconferencing, text messaging, and other communications over landline, cellular, and internet usually are stored by the manufacturer or developer of the device and carriers.

☐ *Signal* is an end-to-end encrypted voice, video, and instant message application for Android, Chrome OS, iOS, macOS, and Windows.

☐ Only download *Signal* from the developer site at *https://signal.org*.

16 Internet Activity

If you go to a coffee shop or at the airport, and you're using open wireless, I would use a VPN service that you could subscribe for 10 bucks a month. Everything is encrypted in an encryption tunnel, so a hacker cannot tamper with your connection.

–Kevin Mitnick[1]

What You Will Learn in This Chapter

- Understand Virtual Private Network (VPN)
- Understand gateway VPN
- Search for a VPN host
- Install and configure VPN
- File share within a Hamachi mesh VPN
- Resolve email conflicts with VPN
- Understand Domain Name System (DNS)
- Test for DNS Leak

What You Will Need in This Chapter

- [Optional] iCloud+ account (paying for more than default 5GB iCloud storage)
- Access to your home or office router and the administrator credentials

---

[1] *https://en.wikipedia.org/wiki/Kevin_Mitnick*

16.1 Virtual Private Network (VPN)

In case you have been sleep-reading through this book, let me repeat my wake-up call: *They are watching you on the Internet. They* may be the automated governmental watchdogs (of your own or another country), government officials (again, of your own or another country), the administrator or owner of your local area network router, bored staff at an Internet Service Provider or broadband provider, a jealous (slightly wackadoodle) ex, high school kids driving by your home or office or sitting on a hill several miles away, marketing groups, or outright professional criminals.

Regardless, your device, data, and privacy are at risk.

Perhaps one of the most important protection steps you can take is to encrypt the entire Internet experience all the way from your computer, through your broadband provider, to a point where your surfing, chat, webcam, email, etc. cannot be tracked or understood. This is accomplished using a technology called *Virtual Private Network[2] (VPN)*.

16.2 Gateway VPN

There are two fundamental flavors of VPN. The most common is called a *gateway VPN*. Mesh VPN is discussed later. Historically, gateway VPN involved the use of a VPN appliance resident at an organization. Telecommuting staff can use the VPN gateway so the Internet acts like a very long Ethernet cable connecting the staff person's computer to the office network. In addition, all data traveling between the user's computer and the gateway is military grade encrypted. The downside to this strategy is that these appliances are expensive (from $600 to several thousand dollars), and they require significant technical experience to properly configure and maintain.

The gateway VPN concept works like this:

---

[2] *http://en.wikipedia.org/wiki/Virtual_private_network*

1. Your device has VPN software installed and configured to connect to a VPN server at the office. This server is connected to your office network.

2. On your device, you open the VPN software and instruct it to connect to the VPN server. This typically requires entering your authentication credentials of username and password.

3. The VPN server authenticates you and begins the connection between itself and your computer. From this point forward, all data traveling between your device and the VPN server are encrypted. Data between the VPN server to the greater internet is not encrypted, unless connecting to a server or service using encryption itself, such as a website using https.

4. Once your data reaches the VPN server, it is forwarded to the appropriate service on your organization's network such as a file server, printer, and mail server.

Although this may sound complex, all a user must do is enter a name, password, and sometimes a key. Everything else is invisible to the user. The only indicator that anything is different is your speed is slower than normal. This is due to the overhead of the encryption/decryption process.

We can use this same strategy to securely surf the Internet by using a VPN service that acts as an intermediary between your device and the Internet.

Using this strategy of a VPN Internet server, all your Internet traffic is military grade encrypted between your computer and the VPN server. It is not possible to decipher your traffic including usernames, passwords, data, or even the type of data coming and going.

One downside is that once the data exits the VPN server, it is readable. However, your data is intermingled with thousands of other user's data, making the process of tweezing out your data a task that only the NSA can accomplish.

Another concern is that some VPN providers maintain user activity logs. This is the law in most countries, so that government agencies can review who is doing what through the VPN. Ideally, you want to work only with a VPN provider operating in a country that does not require logs, and in fact, does not keep logs.

There are thousands of VPN Internet Servers available. Most of them are free. I do not recommend using the free services for two reasons:

- You get what you pay for. Typically, here today, gone tomorrow, unstable, etc.

- You do not know who is listening at the server side of things. Remember, your data is fully encrypted up to the server. But once the data reaches the server on the way to the Internet, it is readable. There must be a high degree of trust for administration of the VPN server.

There are hundreds of legitimate VPN hosts. There are thousands of illegitimate VPN hosts. When determining the best VPN provider for your use, here are key variables to assess:

- **Privacy Jurisdiction.** Must the VPN host comply with US National Security Letters, or other governmental requirements for disclosure of your information? If you care about anybody accessing all your Internet data, the geographical location of the host headquarters is important.

- **Logs**. Are logs kept on client activities? In many countries, it is required by law that all Internet providers maintain logs of client activities. If so, although the logs may not record *what* you were doing, they keep a record of *where* you traveled. It is ideal to have a VPN provider that keeps no logs.

- **Speed**. How fast is your Internet experience? Using VPN introduces a speed penalty due to the encryption/decryption process, as well as the need to process all incoming and outgoing packets through a server instead of point-to-point. VPN providers can reduce this penalty in several ways: faster servers, reduced clients:server ratio, better algorithms, content filtering to remove advertisements and cookies, and faster server internet connections.

- **Support**. VPN adds a layer of complexity to your Internet activities. Should something not work correctly, you do not want to be the one troubleshooting. Ideally, your VPN provider has 24/7/365 chat support. Even better if they offer telephone support.

- **Cross-Platform Support**. Most of us have more than one device. It would be madness to have to use a different VPN product for each of these. Look for a provider that supports all your current and potential devices.

- **Multi-Device Support**. Most, but not all, providers now offer from 3-5 concurrent devices licensing. This allows your VPN service to be operational on all your devices at the same time. Providers that offer only single-device licensing may be quite costly should you have multiple devices.

- **DNS-Leak Protection**. A VPN encrypts all data that comes and goes from your device. But before you reach out to the Internet to connect to your email, a website, or text, your device must connect to a DNS server for guidance on where to find the mail, web, or text server. If you use your default DNS server (typically one from your Internet broadband provider), the data between your system and the DNS server is not encrypted *and* is recorded. It is ideal if your VPN provider offers their own DNS servers. Using this strategy, the data between your device and the DNS server is now encrypted or not logged.

- **VPN Protocol.** There are several network protocols available for encryption. We currently recommend IKEv2. This is the most current, and one of the few protocols with the ability to automatically activate upon accessing the Internet and deactivate when not in use.

- **Pricing.** This is sometimes directly related to the quality of service, and occasionally directly related to the greediness of those running the business. Look for reasonable pricing for the services offered, as well as how many concurrent connections you are allowed. Some hosts allow only 1 connection. Others offer 6 or more, which allows for you to have your computer, phone, tablet connected via VPN, as well as those of a family member.

Now you may be asking yourself: If VPN is so great, why doesn't everyone know about and use it?

Great question! As with everything else in life, there is bad that comes with the good. Each person needs to weigh the pros and cons for their situation. I *always* have VPN active, but I'm *always* doing work! There are three primary downsides to VPN.

- It slows your Internet performance, often by 50% or more. If all I want to do is to stream Netflix to my computer, I'd turn VPN off to reduce the pauses induced by a slow Internet connection.

- If you have selected a VPN server outside of your home country, you may have unintended consequences due to the *Internet* servers thinking you are resident in that other country. For example, Google searches display in the language native to that country. This is considered a feature of the *Proxy Server* function built into VPN and is used in restrictive countries to view news across the border that normally is filtered out by a home country.

- Some websites and services refuse to work with VPN active. This is usually because they want to be able to track you. You will need to decide which you want more – a particular application or privacy.

16.2.1 Assignment: Search for a VPN Host

In this assignment, you search for at least three VPN hosts that meet your needs.

1. Based on the list of criteria listed above, make a list of VPN Host *must-haves* and *prefer-to-haves*.

2. Open a web browser to https://www.safetydetectives.com.

3. From the site menu, select *Best VPN*.

4. From the sidebar, scroll down then select *VPN Comparison Charts*. The URL is https://www.safetydetectives.com/best-vpns/#comparison.

5. There are two charts–*Simple VPN Comparison Chart* and *Detailed VPN Comparison Chart*. For this assignment, select *Simple VPN Comparison Chart*.

6. Tap the *Show* <pop-up menu> *entries*, then select *All*.

7. To sort by your first must-have, tap in that column header. For this assignment, tap on *Privacy Logging*.

8. From the few hosts meeting these minimum requirements, view their websites to evaluate if any meet your needs.

9. Outside this course, you may sign up with one or more of the hosts meeting your needs, put its service through an evaluation, unsubscribe and request a refund if they do not meet with your expectations.

10. Below this *Simple* form is a *Detailed VPN Comparison Chart.*

16.3 NordVPN

One of our favorite VPN providers is *NordVPN*[3]. They offer subscriptions as short as a month, as long as 2 years. With this you get servers in many countries, use on multiple devices, unlimited and highly responsive bandwidth, specialized VPN servers, and a good price.

16.3.1 Assignment: Create a NordVPN Account

In this assignment, you create a NordVPN account. NordVPN has a 30-day guarantee. If within 30 days you do not see value in the service, you may request a refund.

1. To open a 30-day trial account, open a browser then visit *https://free.nordvpn.com/.*

2. Tap the *Start Now* button, then tap *Continue to Payment*.

3. Follow the onscreen instructions to create your account.

4. Open the email from NordVPN, then tap the *Activate Now* button.

5. At the *Set password* page, create a password for your NordVPN account, then tap the *Set Password* button.

6. At the *Welcome to NordVPN* page, tap the *Download* button.

7. Tap the *macOS* button > *NordVPN (IKEv2).*

8. The App Store opens. Tap *Download* to install the NordVPN IKEv2 utility.

---

[3] *https://nordvpn.com*

16.3.2 Assignment: Configure NordVPN

We typically recommend using the IKEv2[4] protocol for VPN, both for its strong encryption and for its automatic activation when accessing the Internet. However, NordVPN has developed its own VPN protocol, NordLynx, which we have found to be just as secure, but even faster than IKEv2.

In this assignment, you configure a VPN connection with NordVPN using the NordLynx protocol.

1. Open the *NordVPN* application, found in the */Applications* folder.

2. At the prompt, enter your NordVPN account name and password.

3. Tap the *Preferences* icon in the top left corner. NordVPN Preferences opens.

4. From the sidebar, select *General.* Configure as shown below.

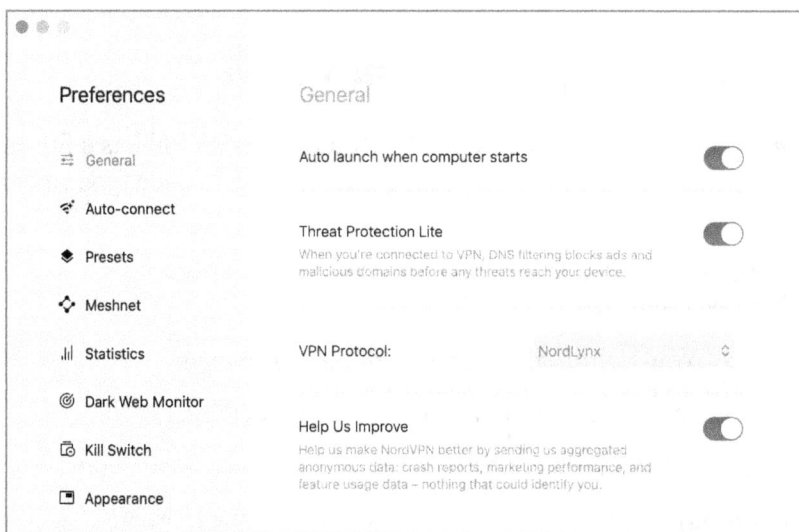

5. From the sidebar, select *Auto-connect.* Configure to your taste. Shown below are my settings, which allows NordVPN to find the best server based on my current location.

---

[4] *https://en.wikipedia.org/wiki/Internet_Key_Exchange*

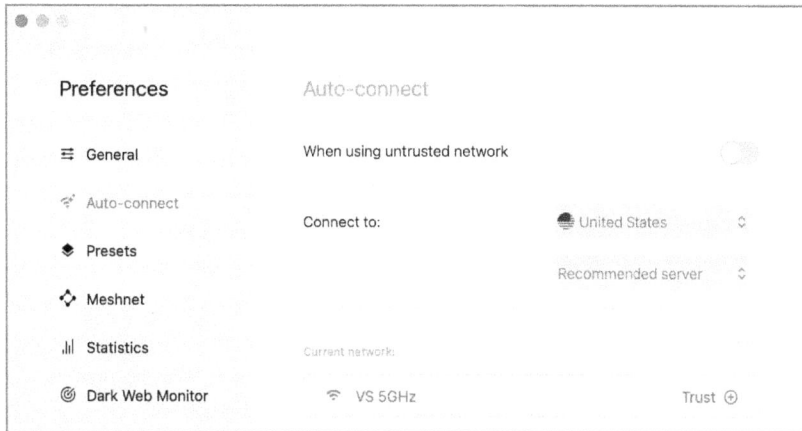

6. From the sidebar, skip *Presets.*

7. From the sidebar, select *Dark Web Monitor,* then *Turn On.*

8. From the sidebar, select *Kill Switch.* For most users this can be left *off.*

9. From the sidebar, select *Select Appearance.* Enable *Both,* to display the NordVPN icon in the menu bar and in the Dock.

10. Close Settings.

11. Close the NordVPN window, but do not Quit the app.

NordVPN interface

There are three places to view your NordVPN activity, each with slightly different controls:

- *NordVPN* app. This can be accessed from the Dock (to open the app), and from the menu bar to connect, disconnect, pause, edit settings, and select a server.

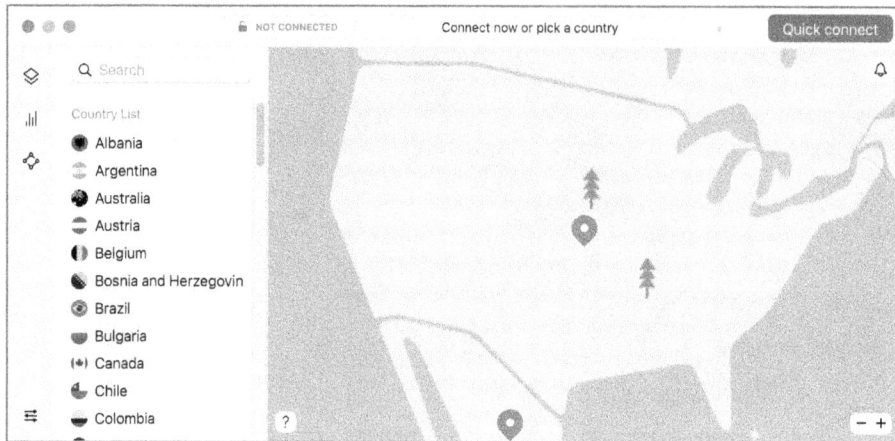

- *System Settings > Network.* There is not much to do here.

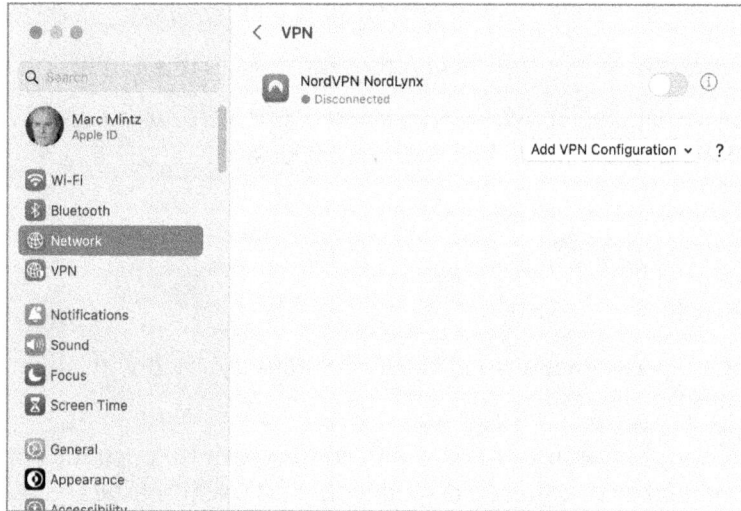

- *VPN* menu icon. If you enable *Show VPN status in menu bar* in the Network System Settings (see above screen shot), you can view and control NordVPN from here as well.

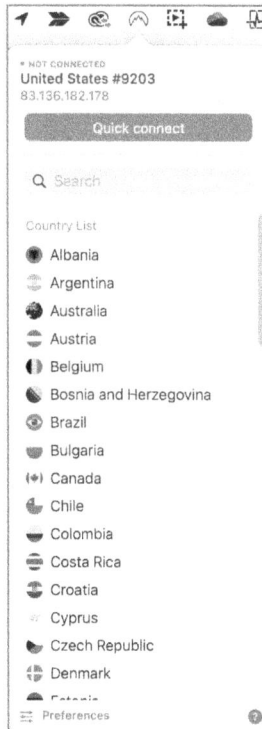

16.4 Resolve Email Conflicts with VPN

Some email servers send up a red flag or block user access to email when the user switches to a VPN connection. This is a good thing as it indicates the email provider is highly sensitive to any possible security breach. In all cases, there is a resolution available although the steps to take vary with each provider.

As an example, I list below what occurs when using VPN with a Gmail account, and how to gain access to your email after the blockage.

1. The user starts a VPN program to encrypt all data between the user's computer and the Internet.

2. The user attempts to receive Gmail.

3. Google sees attempted access from an unknown machine (the Proxy Server), and blocks access to the account.

4. Both an email and a text from Google are sent to notify the user of suspicious activity.

5. Select the link in either the email or text message.

6. The first support file opens. Select the link.

7. In the authentication window, enter your email and password, then select the *Sign In* button.

8. Another support window opens, explaining the next steps to take. Select the *Continue* button.

9. The final support window opens. Following the instructions, return to your email application and access your Gmail within 10 minutes. This provides Google with the authentication to release your account.

16.5 Mesh VPN

Another way in which VPN can be configured is a *mesh VPN*. This strategy places multiple computers within the same virtual network regardless of where they are geographically located on the Internet. All the computers operate as if they are on the same physical network, and all traffic between each of the computers is military grade encrypted. Mesh VPN is ideal for groups of people to exchange files, screen share, and access databases from each other, while maintaining full privacy from the outside world.

NordVPN includes *Meshnet* with their service, allowing up to 10 of your own devices and 60 devices from other NordVPN users to share a virtual network at no additional cost.

If you need to include a greater number of devices, or don't want to use NordVPN, *LogmeIn Hamachi*[5] offers virtual networking with almost unlimited devices.

---

[5] *https://www.vpn.net*

16.6   Domain Name System (DNS)

Most activities on the Internet require pointing to a specific device by use of an address. For example, to use my email, the email software must be able to locate my email server. It does this by looking for *mail.thepracticalparanoid.com.*

Such human-readable names (called a *Fully Qualified Domain Name,* or *FQDN*) work well for you and me, but not so much for computers. Computers expect to use a TCP-IP address. In the case of this server, that is *172.217.3.39.*

The translation from FQDN to TCP-IP address is done by way of the *Domain Name System (DNS).* The process works like this:

1. The user, software, or setting enters the FQDN. For this example, I may enter it in a web browser so that I can view my email.

2. The browser has no idea how to find the FQDN, so it sends the request to the designated DNS server. For macOS, this is configured in *System Settings > Network > Advanced > DNS.*

3. The DNS server maintains a database of all registered FQDN and their TCP-IP address. It sends the search result back to my device, allowing the browser to take me to my email.

The system is fast and stable. The concern is that your Internet provider is usually your DNS provider. This allows the provider to monitor and log most of your Internet activity without your consent or knowledge.

If you use VPN, and your VPN provider has DNS Leak Protection, your Internet provider cannot see your DNS queries. But you may not be using VPN all the time.

To protect your Internet activity from being logged by your Internet provider, manually configure your DNS server to be one that ensures your privacy. I recommend the 1.1.1.1 and 1.0.0.1 servers hosted by Cloudflare[6]. They do not monitor activity, nor do they maintain logs.

In addition to their 1.1.1.1 and 1.0.0.1 DNS servers which focus on security, privacy, and speed, Cloudflare has two additional free DNS servers:

---

[6] *https://www.cloudflare.com/learning/dns/what-is-1.1.1.1/*

- Malware blocking only, 1.1.1.2 and 1.0.0.2
- Malware and Adult Content blocking, 1.1.1.3 and 1.0.0.3

16.6.1 Assignment: DNS Leak Test

If you have a DNS leak, your DNS records (including internet travels) may be visible to others.

In this assignment, you test for a DNS leak.

1. Open a browser to *https://www.dnsleaktest.com*.

2. Select the *Extended Test* button.

3. When the test completes, verify your ISP is not listed, and only your desired DNS host or VPN provider displays.

    o This is the result when using my default setup, without VPN:

    o This is the result when using VPN:

Test complete

Query round	Progress...	Servers found
1	......	1
2	......	1
3	......	1
4	......	1
5	......	1
6	......	1

IP	Hostname	ISP	Country
64.44.80.156	156-80-44-64-.reverse-dns.	Nexeon Technologies	Greenwood Village, United States

16.6.2 Assignment: Secure DNS Traffic

In this assignment, you manually configure your DNS settings to use Cloudflare instead of the default (typically your Internet provider) DNS.

1. Open *Apple* menu > *System Settings*.

2. In the *Search* field, enter DNS.

3. From the sidebar, select *DNS servers*.

4. From the sidebar, select *DNS*. The *DNS Servers* appears in the main window area.

5. Tap the + button, then add 1.1.1.1.

6. Tap the + button, then add 1.0.0.1.

7. Tap the – button, then delete any other DNS servers.

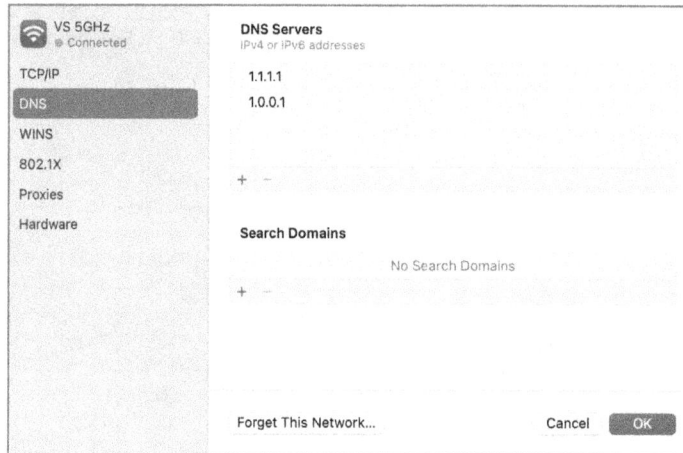

8. Tap the *OK* button.

From now on, all your DNS searches are performed securely by Cloudflare.

16.7 Private Relay

Private Relay[7], new with macOS 12, iOS 15, and iPadOS 15, prevents third-party sites from determining your web-browsing habits, and blocks the ability to fingerprint you. The third-party sees an IP address that is not yours and what that IP address is doing on the site. The same is true for anyone else with access to your internet traffic.

This is done by:

- Encrypting all internet-bound traffic when using Safari

- Routing internet-bound traffic from Safari through two proxy servers–one owned by Apple, the other by another provider–which strip off your IP address and separates your IP from the web activity.

As of this writing, Private Relay is available to all iCloud+ users (paid iCloud) in most but not all countries. Private Relay will apply to:

_____

[7] *https://www.apple.com/newsroom/2021/06/apple-advances-its-privacy-leadership-with-ios-15-ipados-15-macos-monterey-and-watchos-8/*

- All Safari web browsing
- All DNS queries as users enter site names
- All insecure HTTP traffic

Private Relay will not apply to:

- Local Area Network
- Private domain name queries
- Traffic using regular VPN
- Internet traffic using a proxy

Private Relay is different from regular VPN:

Feature	Private Relay	VPN
Hide true IP address	☑	☑
Unblock websites blocked by ISP	☑	☑
Works with all browsers and mail apps Apple states Private Relay only works with Safari. However, as of this writing Private Relay works with all browsers and the Mail.app.	⊘?	☑
Unblock geo restricted content Private Relay gives no option for user to select apparent geo location, but it is always within the same country. VPN allows choosing from many countries.	⊘	☑
Cost Private Relay is free (after paying for iCloud+). Free VPN should never be used.	☑	⊘
Security Private Relay uses a dual-hop architecture, not even Apple can decipher the data. VPN can see your activity locations.	☑	☑?

16.7.1 Assignment: Use Private Relay

In this assignment, you configure Private Relay to automatically activate when using Safari.

- Prerequisite: macOS 12 or higher

- Prerequisite: Safari 15 or higher

- Prerequisite: iCloud+ account. If you are paying for more than the default 5GB of iCloud Drive storage, you have an iCloud+ account.

1. Open *System Settings > Apple ID > iCloud.*

2. Select the *Options* button to the far right of *Private Relay.*

3. Select the *Turn On Private Relay* button.

4. In the *IP Address Location* area, select either *Preserve Approximate Location* or *Use Broader Location*, then tap *OK* button.

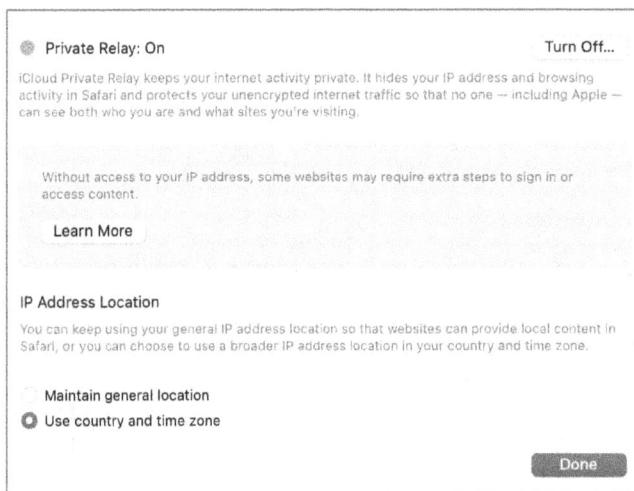

5. Private Relay is now active whenever using Safari.

16.8   Internet Activity Lessons Learned

☐ Virtual Private Network (VPN) encrypts data between your device and the VPN server.

☐ Gateway VPN allows devices to connect with the internet using encryption between the device and the VPN server.

☐ Mesh VPN allows people and devices connected to the internet while geographically separated to function as though all devices are on the same encrypted local network.

☐ Even if using VPN, your internet provider or VPN provider may be able to track your activities by your DNS activity. This is called a DNS leak. The better VPN providers have built-in secure encrypted DNS services.

☐ By configuring your *System Settings > Network > Advanced > DNS* tab to use Cloudflare DNS services with 1.1.1.1 and 1.0.0.1, you always have secure DNS operating.

☐ Some websites and services refuse to work with VPN active. This is usually because they want to be able to track you. You will need to decide which you want more – a particular application or privacy.

☐ Private Relay is an Apple-only feature, available to iCloud+ account holders, using macOS 12, iOS 15, and iPadOS 15, with Safari. It encrypts all internet-bound traffic and separates your IP address from your web activities.

17 Social Media, Apple ID and iCloud

A lot of people say that social media is making us all dumber, but I not think that.
–Unknown author

What You Will Learn in This Chapter

- Protect your privacy on social media
- Harden Facebook security
- Harden LinkedIn security
- Harden Google security
- Harden Apple ID & iCloud security
- Harden Microsoft security

What You Will Need in This Chapter

- No additional resources required.

17.1 What, Me Worry?[1]

Social media[2] has provided us with new levels of connectivity and communication. Victims of natural and man-made catastrophes can instantly assure family and friends of their location and health. Job searches have been

---

[1] *Alfred E. Neuman, https://en.wikipedia.org/wiki/Alfred_E._Neuman*
[2] *https://en.wikipedia.org/wiki/Social_media*

reduced to a few mouse taps. And those who once would have had no voice, may now have a voice that is heard around the world.

Virtually all social media is free to the user. Few users ever question how a service such as Facebook, that may have operating expenses of 12 *billion* dollars[3], not only can afford such expense, but then go on to have a profit of 10 *billion* dollars. The business model for social media is partially based on advertising, but far more on selling information about *you* to the advertisers.

Social media knows more about you than your mother. Their systems know what you are doing on their site–as well as all ancillary sites–including how long you stay on each page, where you came from to land on that page, where you jump off to, what you have purchased, your interests, issues that prompt strong emotion in you, and more. The data and metadata held by social media sites have been shown to be extremely accurate predictors of behavior. It is *this* information that is so very valuable to advertisers, and potential employers, and the government.

It has become the norm for HR departments to scan all social media of a potential employee. The belief is that what a person expresses in social media is a more accurate and honest representation than the employment form or initial interview.

Government agencies closely track social media to predict and stop the next terrorist attack, as well as other lesser crimes.

Then we have criminals tracking social media for mentions of things like "*...our entire family is leaving for a month-long vacation next Friday....*"

17.2 Protect Your Privacy on Social Media

Privacy on social media starts with one step: understand that social media watches *everything* you do, when you do it, how long you do it, and who you do it with. Social media also becomes your *brand*, intentional or not. It colors how others– friends, family, employers, and government–see you. To this effect, you must manage your brand. This is done by being fully mindful of *everything* about your social media pages. Does it represent you in the best light? If taken out of context, how would your data be interpreted?

---

[3] *https://finance.yahoo.com/quote/FB/financials?ltr=1*

The next step is to take whatever measures the social media site offers you to ensure that only those you *want* in, do get in. In the following sections, we show how to protect your security using Facebook, LinkedIn, Google, Apple ID, and iCloud privacy features.

17.3   Facebook

Facebook is the reigning king of social media. Whether due to lax security concerns, or the wild popularity of the site, account breaches and hacks are common. Fortunately, Facebook has taken user-centric security controls out of the shadows and made them easily accessible.

Facebook collects information about you from three different levels:

- Data about your interactions while on Facebook.

- Data about your interactions with websites that have Facebook tracking.

- Data about your interactions with apps that have Facebook tracking.

- According to a *Ghostery*[4] report, Facebook has trackers on 23% of websites.

- As reported in *Slashdot*[5], 61 of the top 100 apps have Facebook tracking.

17.3.1 Assignment: Facebook Security and Login

In this assignment, you secure your Facebook password.

- Prerequisite: Ownership of a Facebook account.

1. Open a browser to *https://facebook.com/*, then log into your account.

2. Tap the downward triangle at the top right corner of the Facebook window. Select *Settings & Privacy > Settings*.

---

[4] *https://www.ghostery.com/tracking-the-trackers-2020-web-trackings-opaque-business-model-of-selling-users/*

[5] *https://tech.slashdot.org/story/21/08/29/1758218/facebook-has-trackers-in-25-of-websites-and-61-of-the-most-popular-apps*

3. From the sidebar, select *Security and Login.* Carefully review and configure each of the areas and settings. Of particular importance:

   a. Are any of the *Where You're Logged In* devices not yours?

   b. Change your password to a strong, unique password.

   c. Enable Two-Factor Authentication (2FA).

4. From the sidebar, select *Privacy.* Carefully review and configure each of the areas and settings.

5. From the sidebar, select *Face Recognition.* This should be disabled for most of us. If enabled, Facebook can use this to track all images with you in them.

6. From the sidebar, select *Profile and Tagging.* Carefully review and configure each of the areas and settings.

7. From the sidebar, select *Public Posts.* Carefully review and configure each of the areas and settings.

8. From the sidebar, select *Blocking.* Carefully review and configure each of the areas and settings.

9. From the sidebar, select *Location.* This should be disabled for most of us.

10. From the sidebar, select *Stories.* This should be disabled for most of us.

11. From the sidebar, select *Apps and Websites.* Carefully review and configure each of the areas and settings.

12. From the sidebar, select *Instant Games.* Games are a huge security vulnerability. I recommend removing any games listed.

13. From the sidebar, select *Business Integration.* Unless you trust and value the service offered here, remove them.

14. From the sidebar, select *Ads > Ad Settings.* Carefully review and configure each area and setting, drilling down the multiple levels you find. This section is critically important to help restrict your Facebook activities from being known by marketeers.

17.3.2 Assignment: What Does Facebook Know About You

Much of the information Facebook has about you may be found in the assignments above. But there is more that can be mined. In this assignment, you have Facebook provide the information it holds and shares about you.

The data that Facebook has on each member includes the following: (For a full list, visit *https://www.facebook.com/help/405183566203254*).

What info is available?	What is it?	Where can I find it?
About Me	Information you added to the About section of your Timeline like relationships, work, education, where you live and more. It includes any updates or changes you made in the past and what is currently in the About section of your Timeline.	Activity Log Downloaded Info
Account Status History	The dates when your account was reactivated, deactivated, disabled, or deleted.	Downloaded Info
Active Sessions	All stored active sessions, including date, time, device, IP address, machine cookie and browser information.	Downloaded Info
Ads Clicked	Dates, times, and titles of ads clicked (limited retention period).	Downloaded Info
Address	Your current address or any past addresses you had on your account.	Downloaded Info
Ad Topics	A list of topics on which you may be targeted based on your stated likes, interests, and other data you put in your Timeline.	Downloaded Info
Alternate Name	Any alternate names you have on your account such as a maiden name or a nickname.	Downloaded Info
Apps	All the apps you have added.	Downloaded Info
Birthday Visibility	How your birthday appears on your Timeline.	Downloaded Info
Chat	A history of the conversations you've had on Facebook Chat (a complete history is available directly from your messages inbox).	Downloaded Info

What info is available?	What is it?	Where can I find it?
Check-ins	The places you've checked into.	Activity Log Downloaded Info
Connections	The people who have liked your Page or Place, RSVPed to your event, installed your app or checked in to your advertised place within 24 hours of viewing or clicking on an ad or Sponsored Story.	Activity Log
Credit Cards	If you make purchases on Facebook (for example: in apps) and have given Facebook your credit card number.	Account Settings
Currency	Your preferred currency on Facebook. If you use Facebook Payments, this will be used to display prices and charge your credit cards.	Downloaded Info
Current City	The city you added to the About section of your Timeline.	Downloaded Info
Date of Birth	The date you added to Birthday in the About section of your Timeline.	Downloaded Info
Deleted Friends	People you've removed as friends.	Downloaded Info
Education	Any information you added to Education field in the About section of your Timeline.	Downloaded Info
Emails	Email addresses added to your account (even those you may have removed).	Downloaded Info
Events	Events you've joined or been invited to.	Activity Log Downloaded Info
Facial Recognition Data	A unique number based on a comparison of the photos you're tagged in. This data helps others tag you in photos.	Downloaded Info
Family	Friends you've indicated are family members.	Downloaded Info
Favorite Quotes	Information you've added to the Favorite Quotes section of the About section of your Timeline.	Downloaded Info
Followers	A list of people who follow you.	Downloaded Info
Following	A list of people you follow.	Activity Log
Friend Requests	Pending sent and received friend requests.	Downloaded Info

What info is available?	What is it?	Where can I find it?
Friends	A list of your friends.	Downloaded Info
Gender	The gender you added to the About section of your Timeline.	Downloaded Info
Groups	A list of groups you belong to on Facebook.	Downloaded Info
Hidden from News Feed	Any friends, apps, or pages you've hidden from your News Feed.	Downloaded Info
Hometown	The place you added to hometown in the About section of your Timeline.	Downloaded Info
IP Addresses	A list of IP addresses where you've logged into your Facebook account (does not include all historical IP addresses as they are deleted according to a retention schedule).	Downloaded Info
Last Location	The last location associated with an update.	Activity Log
Likes on Others' Posts	Posts, photos, or other content you've liked.	Activity Log
Likes on Your Posts from others	Likes received on your own posts, photos, or other content.	Activity Log
Likes on Other Sites	Likes you've made on sites off Facebook.	Activity Log
Linked Accounts	A list of the accounts you've linked to your Facebook account	Account Settings
Locale	The language you've selected to use for Facebook.	Downloaded Info
Logins	IP address, date and time associated with logins to your Facebook account.	Downloaded Info
Logouts	IP address, date and time associated with logouts from your Facebook account.	Downloaded Info
Messages	Messages you've sent and received on Facebook. Note that if you've deleted a message, it will not be included in your download as it has been deleted from your account.	Downloaded Info
Name	The name on your Facebook account.	Downloaded Info

What info is available?	What is it?	Where can I find it?
Name Changes	Any changes you've made to the original name you used when you signed up for Facebook.	Downloaded Info
Networks	Networks (affiliations with schools or workplaces) that you belong to on Facebook.	Downloaded Info
Notes	Any notes you've written and published to your account.	Activity Log
Notification Settings	A list of all your notification preferences and whether you have email and text enabled or disabled for each.	Downloaded Info
Pages You Admin	A list of pages for which you are the admin.	Downloaded Info
Pending Friend Requests	Pending sent and received friend requests.	Downloaded Info
Phone Numbers	Mobile phone numbers you've added to your account, including verified mobile numbers you've added for security purposes.	Downloaded Info
Photos	Photos you've uploaded to your account.	Downloaded Info
Photos Metadata	Any metadata that is transmitted with your uploaded photos.	Downloaded Info
Physical Tokens	Badges you've added to your account.	Downloaded Info
Pokes	A list of who has poked you and who you have poked. Poke content from our mobile poke app is not included because it's only available for a brief time. After the recipient has viewed the content, it's permanently deleted from your system.	Downloaded Info
Political Views	Any information you added to Political Views in the About section of Timeline.	Downloaded Info
Posts by You	Anything you posted to your own Timeline, like photos, videos, and status updates.	Activity Log
Posts by Others	Anything posted to your Timeline by someone else, such as wall posts or links shared on your Timeline by friends.	Activity Log Downloaded Info
Posts to Others	Anything you posted to someone else's Timeline, such as photos, videos, and status updates.	Activity Log

What info is available?	What is it?	Where can I find it?
Privacy Settings	Your privacy settings.	Privacy Settings Downloaded Info
Recent Activities	Actions you've taken and interactions you've recently had.	Activity Log Downloaded Info
Registration Date	The date you joined Facebook.	Activity Log Downloaded Info
Religious Views	The current information you added to Religious Views in the About section of your Timeline.	Downloaded Info
Removed Friends	People you've removed as friends.	Activity Log Downloaded Info
Screen Names	The screen names you've added to your account, and the service they're associated with. You can also see if they're hidden or visible on your account.	Downloaded Info
Searches	Searches you've made on Facebook.	Activity Log
Shares	Content (ex: a news article) you've shared with others on Facebook using the Share button or link.	Activity Log
Spoken Languages	The languages you added to Spoken Languages in the About section of your Timeline.	Downloaded Info
Status Updates	Any status updates you've posted.	Activity Log Downloaded Info
Work	Any current information you've added to Work in the About section of your Timeline.	Downloaded Info
Vanity URL	Your Facebook URL (ex: username or vanity for your account).	Visible in your Timeline URL
Videos	Videos you've posted to your Timeline.	Activity Log Downloaded Info

1. Open a web browser, then log into your Facebook account.

2. In the top right corner, tap the *triangle* icon > *Settings* > *Your Facebook Information* in the sidebar.

3. Select *Download a Copy of Your Information*.

4. In the *Download Your Information* page, in the *Date Range* field, select *All of my data,* then tap *Create File.*

    *   It may take several hours for your file to be ready. Facebook sends you an email notifying you when it is ready, along with a download link.

5. In the *Download Your Information* page, select the *Start My Archive* button.

6. At the authentication prompt, enter your Facebook password, then select the *Submit* button.

7. Select the *Start My Archive* button.

8. Your request is submitted to Facebook. Depending on the size of the database, it may take a day or more for you to receive an email link to download your data. Select the *OK* button.

9. Watch for your Facebook link in your email. When the link arrives, tap the link to access your archive.

Manage your Activity Log

10. From the sidebar, select *Your Facebook Information > Activity Log.*

11. The sidebar displays all your recent history.

12. To delete records of specific activities, tap on the activity, tap on the ... button, then select *Delete.*

Manage your off-Facebook Activity Log

13. From the sidebar, select *Your Facebook Information > Off-Facebook Activity.*

14. Select *Manage Your off-Facebook Activity.* A list displays of recent activity that other businesses and organizations shared with Facebook.

15. If you do not want an organization you met on Facebook to share information that organization has gathered about you with Facebook, notify the organization to not share your information by tapping: *Turn off future activity from <company name>.*

16. Exit from Facebook *Settings.*

17.3.3 Assignment: What Does Off-Facebook Know About You

Websites and apps that have Facebook tracking installed will feed your activities on the sites and apps to Facebook. You can view which sites and apps are involved, and the ability to manage this information. More detailed information may be found here.[6]

In this assignment, you view your off-Facebook tracking.

View Off-Facebook Activity

1. Open a browser then log into your Facebook account.

2. Select your *Profile Picture > Settings & Privacy > Settings*.

3. In the *Settings* page, select *Your Facebook Information* from the sidebar > *Off-Facebook activity > View*.

4. The *off-Facebook activity* page opens.

5. Select *Recent Activity* to view where your Facebook data has been.

6. For this assignment, close the *Activity Details* window.

17.4   LinkedIn

While Facebook is the current reigning king of non-business social media, LinkedIn holds the crown for business. Whether it be to market one's services, look for a new job, or simply network with other businesspeople, LinkedIn is the place to be.

Just as with all other social media sites, it is vital to be mindful of privacy and security on LinkedIn. It could be business suicide to have anything but the very best information associated with your account.

---

[6] *https://www.facebook.com/help/2207256696182627*

17.4.1 Assignment: LinkedIn Account Security

The need for a strong LinkedIn password is no different than for your computer.

In this assignment, you change your current LinkedIn password to a strong password.

- Prerequisite: Access to a LinkedIn account.

1. Open a browser to *https://linkedin.com/*.

2. From the tool bar, select the *Me* icon > *Settings & Privacy*.

3. The *How LinkedIn uses your data* page opens.

4. From the sidebar, select *Account Preferences*. The *Account Preferences > Profile Information* page opens.

5. One by one, select each section starting with *Name, location, and industry,* to verify the information is accurate, places you in the best light, and is shared only with those who should see this information.

6. Repeat steps 4 and 5 for each sidebar section.

17.4.2 Assignment: Find What LinkedIn Knows About You

In this assignment, you download all the data that LinkedIn (admits) to knowing about you.

1. Open a web browser, then log into your LinkedIn account.

2. From the *Me* menu in the top right of the page, select *Settings & Privacy*.

3. From the sidebar, select *Data Privacy*. The *How LinkedIn uses your data* page opens.

4. Scroll down to tap *Export your data*.

5. In the *Get a copy of your data* area, select *Download larger data archive*, then tap *Request archive* button. You are notified by email when the report is available for download:

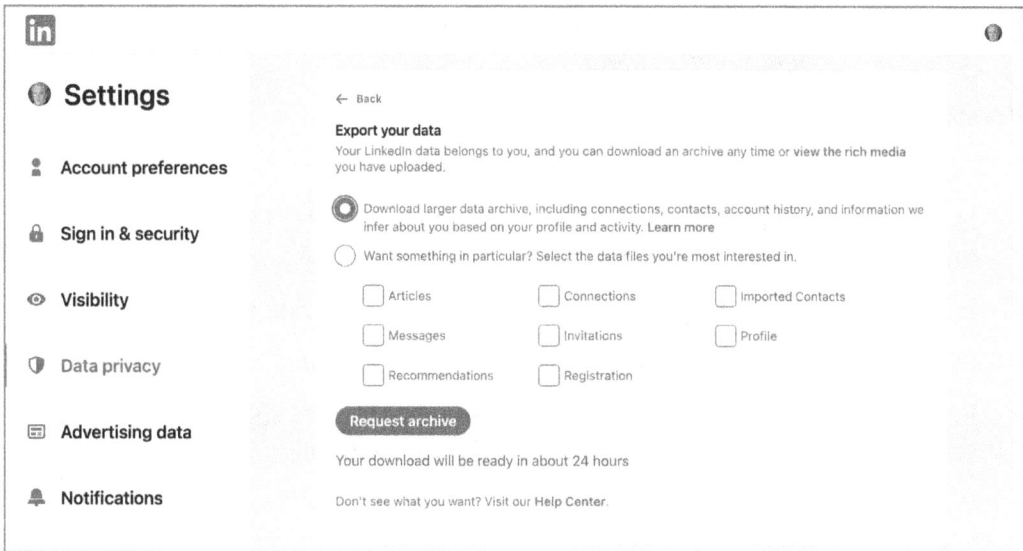

17.5 Google

Although most people think of Google as a search engine, it has become far more than that. Over one billion people use Google mail, maps, YouTube, and Google Play services[7]. Chances are you use Google every day. The result is a tremendous warehouse of data points about your searches, site visits, hangout partners, purchases, and so much more.

And yet, at the same time Google provides the tools to help guard your privacy. NOW is a good time to put these tools to use.

17.5.1 Assignment: Google Sign-In & Security

In this assignment, you begin the process of securing a Google account.

• Prerequisite: Access to an existing Google account.

---

[7] *http://www.digitaltrends.com/web/gmail-joins-the-billion-users-club/*

1. Open a web browser to sign in at the Google Security page *https://myaccount.google.com.*

Personal Info

2. From the tab or sidebar, select *Personal Info.*

3. In the *Basic Info* area, verify all information is accurate. This is also a good time to change your Google password to be more secure.

4. In the *Contact info* area, verify all information is accurate and complete.

5. In the *Choose what others see* area, tap *Go to About Me.*

6. In the *About Me* page, verify all information is accurate and complete, and is viewable only by those who should see the information.

7. When complete, go *Back.*

Data & Personalization

8. From the sidebar, select *Data & Privacy.*

9. Scroll to *Privacy suggestions available > Review suggestions.* Follow the on-screen instructions to improve your privacy, then go *Back.*

10. Scroll down to *History Settings.* Configure *Web & App Activity, Location History,* and *YouTube History* to your taste. I recommend *Pausing* each, to prevent Google building a history on your activity.

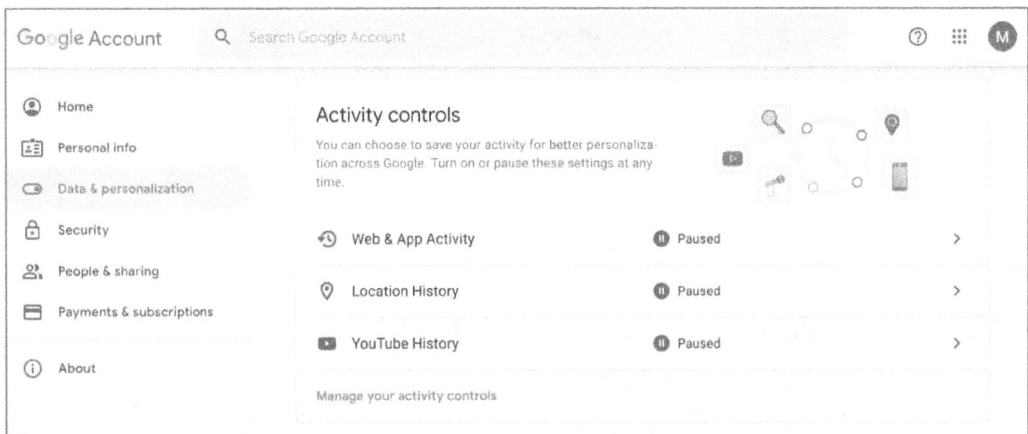

11. In the *Personalized ads* area, turn *off* the following: *My Ad Center,*

12. In the *Info you can share with others* area, configure to your taste, remembering that the less you share, the more secure your data.

13. Tap *Delete a Google service,* authenticate, then review all Google services in use. If a service is active that you no longer use, *delete* it, then go *Back.*

Security

14. From the sidebar, select *Security.*

15. In the *You have security recommendations,* tap *Protect your account.* Review the recommendations, configure to your taste. Then go *Back.*

16. In the *Signing into Google* area, verify you have *2-Stepp Verification* active.

17. In the *Ways we can verify it's you.* Verify your *Recovery phone* and *Recovery email* are accurate and current.

18. Scroll down to *Your devices.*

19. If any of the displayed devices are no longer in use by you, tap *Manage devices, Sign Out* of the device. Then go *Back.*

20. In the *Third-party apps with account access*, if you no longer use any of these apps, tap *Manage third-party access,* select the target app, tap *Remove Access.* Then go *Back.*

21. Scroll down to *Signing into other sites.*

22. In the *Signing in to other sites* area, tap *Signing in with Google.* if you find any apps or services using your Google account to sign in (which means they have access to your Google activities), consider removing them to force them to log in directly. Then go *Back.*

23. Select *Password Manager.* If there are any passwords stored in Google that you do not want there, remove them. Then go *Back.*

24. Select *Linked Accounts.* If you have any linked accounts which give Google access to your data, you may manage that linkage here. Then go *Back.*

People & Sharing

25. From the sidebar, select *People & sharing > Choose what others see > About me.* Configure to your taste, being aware of what information is being made to the world. Then go *Back.*

26. Select *Current settings.* Configure to your taste, being aware of what information is being made to the world. Then go *Back.*

27. Select *Location sharing > Manage location sharing.* Verify you are not sharing your location. Then go *Back.*

17.5.2 Assignment: Find What Google Knows About You with Takeout

In this assignment, you discover what Google knows about you. Google provides a service known as *Takeout* that allows you to download anything and everything Google knows about you (what they admit to knowing about you). When using Takeout, Google delivers to you this data in .zip files. If your history with Google is extensive, the data files are large.

1. Open a web browser, then visit Google Takeout at *https://takeout.google.com/.*

2. From the list of available data subjects, select what you wish to access. I recommend selecting everything. Doing so not only gives you knowledge of what Google knows about you but also provides a backup of all your Google Account data.

3. Scroll to the bottom, then tap the *Next step* button.

4. In the *Google Takeout Create a New Export* page, configure how you wish to receive your *Takeout.* Then select *Create export.*

5. You receive an email when your archives are available.

6. Once all archives are downloaded, double tap to open each. You may find there are duplicate folders at the root level (such as *Google Drive*). Combine the contents of these duplicates.

7. Have fun learning about yourself!

17.6 Microsoft and Windows Security

In 2013[8] Microsoft's built in Security Essentials app failed certification.

In 2015[9], after multiple patches and official sounding remedies were released, it failed again and truly earned its perennial place, positioned on the bottom of all available anti-virus programs.

In 2016[10], Microsoft inadvertently published a Secure Boot "golden key" policy which allowed for anyone to load self-signed or unsigned binaries onto Windows devices. Rumor had it that these were law enforcement backdoors, built into Windows because Microsoft was not willing to fight the FBI over such things. On the bright side, this event gave Apple an easy win in their fight with the FBI about creating such keys in macOS and iOS.

Skip ahead to 2021[11] and it appears that MS has gotten everyone with an outlook exchange account hacked. So, please believe me when I say, you cannot leave Windows 11 security to Microsoft Developers. You must take care of your data, accounts, and access yourself.

Remember that every password and security measure can be broken. Your defense is to make it so difficult and time consuming to break that the hacker moves on to an easier target. Also, most security questions can be accurately guessed or broken through social engineering (*What is your birthday? In what city did your parents marry? What is the name of your first pet?* etc.) Both types of security are based on what you know. And if there is something that you know, someone else can know it as well.

---

[8] *https://www.theverge.com/2013/1/17/3885962/microsoft-security-essentials-fails-anti-virus-certification-test*

[9] *https://redmondmag.com/articles/2015/01/27/security-essentials-fails-antivirus-test.aspx*

[10] *https://www.zdnet.com/article/microsoft-secure-boot-key-debacle-causes-security-panic/*

[11] *https://krebsonsecurity.com/2021/03/at-least-30000-u-s-organizations-newly-hacked-via-holes-in-microsofts-email-software/*

- Note: As of August 2016, NIST has stopped recommending Two-Factor Verification that involves SMS/text messaging as the second factor[12]. This is due to the ease of which this can be intercepted. However, *any* verification that is received via cellular signal (SMS, voice, etc.) is subject to the same vulnerabilities. Currently, the best solution is to use a digital token (a keychain-sized device that displays random number strings), or an *Authenticator* app, which serves the same purpose.

17.6.1 Assignment: Enable Microsoft 2-Factor Authentication

In this assignment, you enable 2FA for your Microsoft account.

- Prerequisite: Access to a Microsoft account.

- Prerequisite: Access to a smartphone.

1. Open a browser to *https://account.microsoft.com.*

2. Sign in with your Microsoft credentials.

3. Tap your avatar in the upper right corner, then tap *My Microsoft Account.*

4. Tap *Security* in the top menu.

5. Tap *Advanced Security Options.*

6. In the *Two-step verification* area, tap to *Turn on.*

7. Verify all options are *Up to date.*

8. Verify *Two-step verification* is *On.*

---

[12] *https://pages.nist.gov/800-63-3/sp800-63-3.html*

9. Scroll down to the *Recovery code* area. Make sure to copy your recovery code, print it, and put it with your other account codes somewhere safe.

17.6.2 [Optional] Assignment: Remove a Device from Your Microsoft Account

All the devices (computers, tablets, and phones that you have ever signed into your MS account from, may still have access to your account, presenting a security vulnerability.

To prevent this from happening, you must remove the device from your Microsoft Account device list.

In this assignment, you remove a device from your Microsoft device list.

1. Open a browser to *https://account.microsoft.com/*.

2. Log in using your MS Account credentials.

3. In the *Devices* area, tap *View all devices*.

4. Go through all your devices to make sure they all belong to you and that you have possession of them. *Remove* all others.

17.7 Apple ID and iCloud Security

In 2012 a well-known journalist had his Apple ID hacked, allowing the hacker full access to the victim's Apple ID, and through that, his iCloud account, including calendar, contacts, and email. This was accomplished not by traditional black hat hacking, but with a bit of social engineering. All the hacker needed was to discover the victim's birthdate and email address associated with his Apple ID. The hacker would look at an email to Apple that said something like, *I've forgotten my Apple ID password and would like to reset it. Here is my birthdate and my email address*. Using this information, the hacker could reset the Apple ID password. Once done, they could access the victim's iCloud website as if they were the actual victim.

I have had several clients whose Music.app/iTunes accounts have been compromised in a similar fashion. One was billed $1,400 in music purchases.

As of macOS 10.13, Two Factor Authentication (2FA) is mandatory for your Apple ID. Adding this security layer makes it extremely difficult for anyone to hijack your Apple ID and make fraudulent purchases.

Remember that every password and security measure can be broken. Your defense is to make it so difficult and time consuming to break that the hacker moves to an easier target. Also, most security questions can be accurately guessed or broken through social engineering: *What is your birthday? In what city did your parents marry? What is the name of your first pet?* etc. When security questions are based on something that you know, someone else can know that.

With Apple 2FA a code is sent to your previously verified devices whenever you sign into your Apple ID on the web to manage your account, purchase something from iTunes, App Store, or iBooks Store from a new (unknown) device or attempt to get Apple ID-related support from Apple. You are prompted to provide this 2FA code before the purchase or support can occur.

If your device has been stolen or lost, log into *https://appleid.apple.com* to remove that device from the verified device list, so that no 2FA code is sent to that device.

- Note: As of August 2016, NIST has stopped recommending Two-Factor Verification that involves SMS/text messaging as the second factor[13]. This is due to how easy it is to intercept the code. But *any* verification that is received via cellular signal (SMS, voice, etc.) is subject to the same vulnerabilities. Currently, the best solution is to use a digital token (a keychain-sized device that displays random number strings), or an *Authenticator* app, which serves the same purpose. At this time, Apple does not use digital tokens or an authenticator app but can provide verification codes to your phone.

17.7.1 Assignment: Enable Apple 2-Factor Authentication

Apple requires 2FA for all accounts. If you upgraded from an older macOS version that did not require 2FA, this assignment will enable it:

---

[13] *https://pages.nist.gov/800-63-3/sp800-63-3.html*

1. Have your smartphone handy. It is required at step 6.

2. Open Safari and go to *https://appleid.apple.com*.

3. Sign in with your Apple ID and password.

4. In the *Security* section, select the *Edit* button.

5. In the Two-Factor Authentication area, select *Turn On Two-Factor Authentication.*

6. Follow the brief on-screen instructions.

17.7.2 [Optional] Assignment: Remove a Device from Apple Two-Factor Authentication

All devices (computers, iPads, iPhones, Apple TVs, etc.) on which you have signed into your Apple account receive *Apple ID Verification Codes* when an attempt is made to access your Apple account. Should one of your devices become lost, stolen, sold, or given to someone, the person who takes possession of it may be able to see your verification codes, presenting a security vulnerability.

To prevent this from happening, you must remove the device from your Apple ID device list.

In this assignment, you remove a device from your Apple ID device list.

• Note: Unless you really do wish to remove the device, skip this assignment.

5. Open a browser to *https://appleid.apple.com*.

6. At the prompt, enter your Apple ID email address and Apple ID password.

7. If you have set up 2-Factor Authentication, your devices display an alert that someone is attempting to access your account. Select *OK*. An *Apple ID Verification Code* appears.

8. Enter the *Apple ID Verification Code* in the alert window in your browser.

9. Scroll down to the *Devices* area.

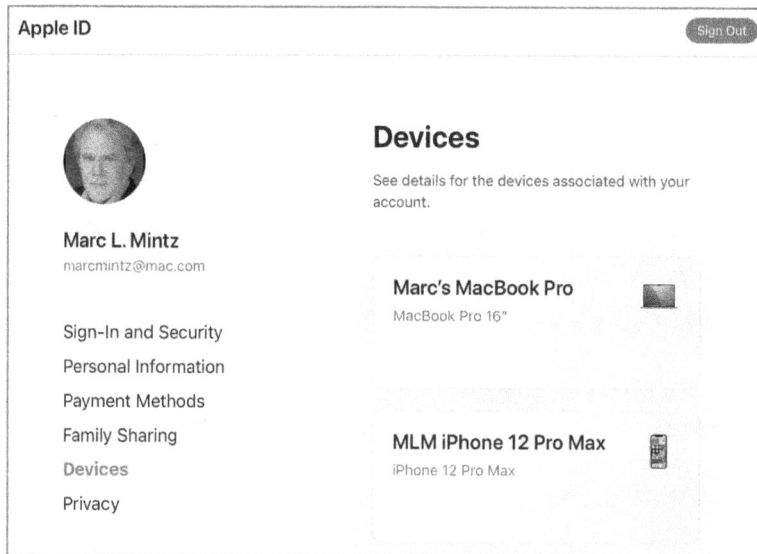

10. Tap on the device to be removed.

11. A pop-up window appears. Tap *Remove from account.*

12. The device is detached from your Apple ID and no longer receives Apple ID Verification Codes.

17.8   Apple Account Recovery and Legacy Contact

New with iOS 15 and macOS 12 is *Apple Account Recovery.* In the event you have lost the ability to access your iCloud data, Account Recovery helps you regain access when you can't reset your password. Some data cannot be recovered, such as keychain, Screen Time, and Health.

Also new with iOS 15 and macOS 12 and associated with Account Recovery is *Legacy Contact.* With Legacy Contact, you can specify someone of trust to have access to data in your account after your death.

17.8.1 Assignment: Account Recovery and Legacy Contact

In this assignment, you set up account recovery, so that in the event you cannot access your Apple account and cannot change its password, you can still gain access.

- Prerequisite: All devices signed in with your Apple ID must be on the current OS (macOS 12 and higher, iOS 15 and higher, iPadOS 15 and higher).

Trusted Phone Numbers

1. Open *System Settings > Apple ID > Password & Security*. The Password & Security widow opens.

2. In the *Trusted Phone Numbers* area, add the phone number(s) that can be used to verify your identity when signing in on a different device.

Account Recovery

3. Tap the Account Recovery *Manage...* button. The *Account Recovery* window opens:

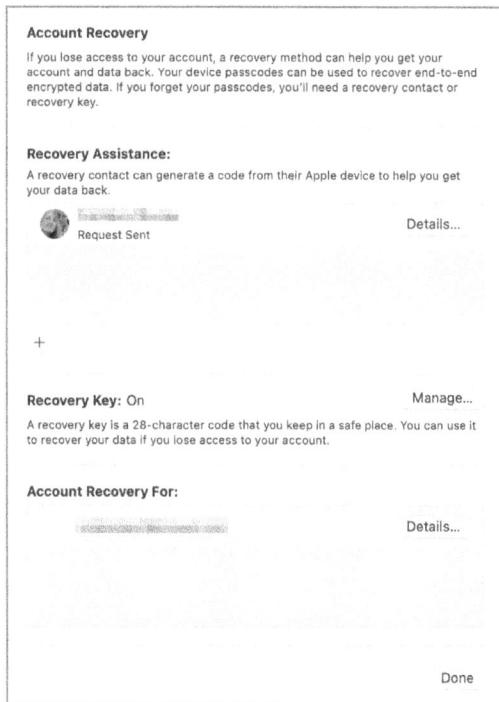

Account Recovery

If you lose access to your account, a recovery method can help you get your account and data back. Your device passcodes can be used to recover end-to-end encrypted data. If you forget your passcodes, you'll need a recovery contact or recovery key.

Recovery Assistance:

A recovery contact can generate a code from their Apple device to help you get your data back.

Request Sent   Details...

+

Recovery Key: On   Manage...

A recovery key is a 28-character code that you keep in a safe place. You can use it to recover your data if you lose access to your account.

Account Recovery For:

Details...

Done

4. Tap the + button, then follow the onscreen instructions to add contact information for someone you trust.

5. Select the Recovery Key *Manage* button, then follow the onscreen instructions to create a recovery key.

6. When complete, tap the *Done* button.

Legacy Contact

7. Select the Legacy Contact *Manage* button, then follow the onscreen instructions to add contact information for someone to access your account after your death.

8. When complete, tap the *Done* button.

Automatic Verification

9. Enable *Automatic Verification* to help bypass CAPTCHA's.

10. Close System Settings.

Trusted Phone Numbers

Used to verify your identity when signing in on a different device or browser.

11. In the *Trusted phone numbers* field, tap *Edit*.

12. Enter at least one phone number that can send you text or voice messages, then tap *Done*.

17.9   Advanced Data Protection

Apple introduced Advanced Data Protection[14] as of iOS 16.2, iPadOS 16.2, and macOS 13.1. This is an option that provides for the highest level of security for iCloud data by enabling end-to-end data encryption between your device and Apples iCloud servers with the encryption keys stored only on your trusted devices and encrypting 23 data categories. Once enabled, not even Apple (or state

---

[14] *https://support.apple.com/en-us/HT202303*

actors, cyber-criminals, or even your government armed with search warrants) can read your data on iCloud.

Prior to the introduction of Advanced Data Protection, or with Advanced Data Protection disabled, the data you store on iCloud has *Standard Data Protection.* 14 data categories of iCloud data are end-to-end encrypted, but the keys are stored on Apple servers. This protects your data from standard hacking attacks but does not protect against targeted attacks or search warrants.

Data Category	Standard Data Protection		Advanced Data Protection	
	Encryption	Key Storage	Encryption	Key Storage
Apple Card Transactions	End-to-End	Trusted Devices	End-to-End	Trusted Devices
Contacts	In Transit & On Server	Apple	In Transit & On Server	Apple
Calendars	In Transit & On Server	Apple	In Transit & On Server	Apple
Health Data	End-to-End	Trusted Devices	End-to-End	Trusted Devices
Home Data	End-to-End	Trusted Devices	End-to-End	Trusted Devices
iCloud Backup (including device and Messages backup)	In Transit & On Server	Apple	In Transit & On Server	Trusted Devices
iCloud Drive	In Transit & On Server	Apple	In Transit & On Server	Trusted Devices
iCloud Mail	End-to-End	Apple	In Transit & On Server	Apple
Maps	End-to-End	Trusted Devices	End-to-End	Trusted Devices
Memoji	End-to-End	Trusted Devices	End-to-End	Trusted Devices
Messaged in iCloud	End-to-End	Trusted Devices	End-to-End	Trusted Devices
Notes	In Transit & On Server	Apple	End-to-End	Trusted Devices
Passwords & Keychain	End-to-End	Trusted Devices	End-to-End	Trusted Devices

Data Category	Standard Data Protection		Advanced Data Protection	
	Encryption	Key Storage	Encryption	Key Storage
Payment Information	End-to-End	Trusted Devices	End-to-End	Trusted Devices
Photos	In Transit & On Server	Trusted Devices	End-to-End	Trusted Devices
QuickType Keyboard Learned Vocabulary	End-to-end	Trusted Devices	End-to-End	Trusted Devices
Reminders	In Transit & On Server	Apple	End-to-End	Trusted Devices
Safari	End-to-End	Trusted Devices	End-to-End	Trusted Devices
Safari Bookmarks	In Transit & On Server	Apple	End-to-End	Trusted Devices
Screen Time	End-to-End	Trusted Devices	End-to-End	Trusted Devices
Siri Information	End-to-End	Trusted Devices	End-to-End	Trusted Devices
Siri Shortcuts	In Transit & On Server	Apple	End-to-End	Trusted Devices
Voice Memos	In Transit & On Server	Apple	End-to-End	Trusted Devices
Wallet Passes	In Transit & On Server	Apple	End-to-End	Trusted Devices
Wi-Fi Passwords	End-to-End	Trusted Devices	End-to-End	Trusted Devices
W1 and H1 Bluetooth Keys	End-to-End	Trusted Devices	End-to-End	Trusted Devices

17.9.1 Assignment: Enable Advanced Data Protection

In this assignment, you enable Advanced Data Protection.

1. Verify that all of your trusted devices are updated to at least iOS 16.2, iPadOS 16.2, or macOS 13.1

2. Open *System Settings* > (User Name) *Apple ID* > *iCloud* > *Advanced Data Protection.*

3. In the *Advanced Data Protection* pane, scroll down to select *Account Recovery* > *Set Up* button.

4. In the *Account Recovery* pane, if you do not yet have *Recovery Assistance* and *Recovery Key* sections completed, do so now, and then tap the *Done* button.

5. Back at the *Advanced Data Protection* pane, tap the *Turn On* button.

6. In the *You will be responsible for your data recovery* pane, tap the *Review Recovery Methods* button.

7. At the prompt, enter your 28-character recovery key, then tap the *Continue* button.

8. At the prompt, enter the login password you use for this device.

9. At the *Advanced Data Protection is On* pane, tap the *Done* button.

10. Tap the *Done* button.

11. In the *iCloud* pane, scroll to the bottom to the *Access iCloud Data on the Web* field.

12. If you would like to be able to access your calendar, contacts, documents, mail, photos, notes, and reminders via a browser from your trusted devices, enable this field.

13. Exit *System Settings.*

17.10 Social Media, Apple ID & iCloud Lessons Learned

☐ Facebook reported a 2020 profit of 12 million dollars, most of this using marketing information about you and other users.

☐ All social media sites should be hardened with 2FA and regular reviews of all security and privacy settings.

☐ You can use Google Takeout to automatically send you on a regular basis a comprehensive report of all it knows about you.

☐ By default, all devices registered with the same Apple ID receive 2FA Apple ID Verification Codes.

☐ New with macOS 12 and iOS 15 are *Apple Account Recovery* making it easier to regain control over your account once locked out, and *Legacy Contact,* where you can grant someone access to your account and digital assets following your death.

☐ New with macOS 13.1 and iOS 16.2 is *Advanced Data Protection*, expanding end-to-end encryption from14 to 23 data categories, and moving the encryption keys from Apples servers to your own trusted devices.

18 Apple Pay and Credit Cards

While money can't buy happiness, it certainly lets you choose your own form of misery.

–Groucho Marx[1]

What You Will Learn in This Chapter

- Escape from the epidemic of credit card theft
- How to configure your computer Apple Pay
- How to use Apple Pay

What You Will Need in This Chapter

- [Optional] Apple Pay compatible debit, credit, or gift card.

18.1   The Epidemic of Credit Card Theft

In 2018, over $24 billion was lost due to card fraud worldwide. How can there be as many as 2 credit card breaches per adult in a year?

One answer is that merchants *love* to keep your credit card information in their greedy little hands. By storing your card information, it is effortless for the merchant to close a sale. With the credit card on record, the buyer has neither the time nor bother of having to search for the card while making a purchase.

---

[1] *https://en.wikipedia.org/wiki/Groucho_Marx*

This arrangement makes the merchant customer credit card database look like Fort Knox to a cyber thief. With enough resources and time, the thief can breach the database, harvest millions of personal identity and credit card records, then turn Target, Home Depot, or any other merchant into their money machine!

There are strategies available to help avoid such problems. A key strategy is to primarily keep pieces of personal and card information on different servers. But even with such a strategy it is possible to breach all the data given adequate resources, especially if an insider is involved.

According to cyber security experts, the real answer lies in preventing the merchant from storing your data in a manner useable by anyone but the individual owner. Several organizations have come up with such a solution in the past few years (Apple Pay, Google Pay, and Samsung Pay). Apple is the first to do it right in the form of *Apple Pay*[2].

When using Apple Pay, the merchant never has access to your credit card number, expiration date, security code, or any other identifiable aspect of your card. All the merchant gets is a one-time-use code that confirms that you do indeed have a valid credit/debit/gift card, and that you (or at least the thumb used to press on the home button) are the rightful holder of said card. The merchant is authorized to charge to that card. But all the merchant has at the end of the transaction is data about the service/merchandise that was purchased, and a one-time-use code! There is nothing in the database worth stealing, and your card information remains hidden behind hardware encryption on your iOS device.

If you do not already have an Apple Pay compatible card, for the sake of security, it may be time to get one. A list of all Apple Pay compatible cards is available at: *https://www.creditcards.com/apple-pay.php*.

Now that you have your card, let's set up your device to use it with Apple Pay.

18.1.1 [Optional] Assignment: Configure Your Device for Apple Pay

In this assignment, you configure your Mac to use a credit card.

- Prerequisite: Possession of an Apple Pay compatible credit, debit, or pre-paid card.

---

[2] *https://en.wikipedia.org/wiki/Apple_Pay*

- Prerequisite: A Mac model with built-in Touch ID.

1. Pull out your bright and shiny debt-making sliver of plastic. You need to enter information regarding the card into your Apple Pay account.

2. Open *Apple* menu > *System Settings* > *Wallet & Apple Pay.*

3. Tap *Add Card* button.

4. Follow the onscreen instructions to add the new card.

5. Tap *Next.* Your card issuer verifies your information and Apple Pay eligibility.

6. When the card issuer has verified the card, tap *Next.*

7. Start using your Apple Pay.

18.1.2 [Optional] Assignment: Use Apple Pay with Online Stores

In this assignment, you make a purchase using Apple Pay.

- Prerequisite: Completion of the previous assignment.

- Prerequisite: Using a Mac with built-in Touch ID.

- Prerequisite: Using Safari. Apple Pay only works with Safari.

1. Go to a brick-and-mortar store that accepts Apple Pay. Close your eyes, take a few steps, you are bound to trip over one. Or if you want to play it safe, look for the Apple Pay logos:

2. Tap the Apple Pay button.

3. Verify that your billing, shipping, and contact information is accurate.

    - If you wish to pay with a different card, tap the pop-up menu icon to the right of your credit card, then select the desired card from your Apple Pay list.

- If you need to enter billing, shipping, or contact information, Apple Pay remembers so you do not need to enter it again.

4. Confirm your purchase.

Ahhhh… Didn't that feel wonderful? Retail therapy *and* no chance of identity or credit card theft.

18.2 Apple Pay and Credit Cards Lessons Learned

☐ Measuring by dollars lost, most credit card theft is due to vulnerabilities at the merchant and credit card processing groups, due to their desire to maintain full records of credit purchases, including all card information.

☐ Currently the best solution to credit card theft and fraud is to prevent the merchant from having access to credit card data. This has been implemented with Apple Pay, Google Pay, and Samsung Pay.

19 [Android and iOS] Medical and SOS Emergency

19.1 [Android and iOS] Medical Emergency

19.1.1 [Android and iOS] Assignment: Configure Health

19.1.2 [Android and iOS] Assignment: Test Health

19.2 [Android and iOS] Emergency SOS

19.2.1 [Android and iOS] Assignment: Configure Emergency SOS

19.2.2 [Android and iOS] Assignment: Test Emergency SOS

19.3 [Android and iOS] Medical and SOS Emergency Lessons Learned

20 When it is Time to Say Goodbye

Don't cry because it's over. Smile because it happened.

–Dr. Seuss[1]

What You Will Learn in This Chapter

- Prepare a computer for sale or disposal
- Remove Firmware password
- Secure erase a storage device
- Reinstall macOS

What You Will Need in This Chapter

- [Optional] Online purchase of *Parted Magic* ($11-$39).

20.1 Prepare a Computer for Sale or Disposal

The time comes when all good things must come to an end. This is just as true for your beloved Macintosh. But your computer holds all your documents, passwords, pictures, web browsing history, and other items you do not want someone else to see. Even if you are tossing your damaged device in the trash, there is the real probability that someone can find it, remove the drive, and harvest all your data.

Your Mac is connected with the Apple environment on several layers, all of which must be disconnected. These include:

---

[1] *https://en.wikipedia.org/wiki/Dr._Seuss*

- Music.app/iTunes
- iCloud
- Messages

Once your mac is disconnected from the Apple environment, the drive itself must be securely erased or destroyed. If the computer will be reused, macOS should be reinstalled.

20.1.1 [Optional] Assignment: Prepare Your Mac for Sale or Disposal

In this assignment, you prepare your Mac to be sold, given away, or discarded.

- **Warning**: Completing this assignment erases your computer. Do not do this assignment unless you intend to permanently erase all content.

Create a Backup

1. As your computer will be securely erased, it is vital to create a secure backup of the drive. As Murphy lives in storage devices, I recommend a minimum of two backups. My preference is for an encrypted Time Machine backup, as well as an internet-based backup with CrashPlan Pro, Backblaze, Carbonite, or SpiderOak. Please see chapter 2 Data Loss for step-by-step instructions.

Sign out of Music

Music.app/iTunes allows only a limited number of devices to be registered to any one Apple ID. If you do not sign out of Music, it thinks this device is still registered with your account, possibly preventing you from adding a new device.

2. On the target computer to which you are saying goodbye, open the *Music* app.

3. From *Music* app, select *Account* menu > *Authorizations* > *Deauthorize This Computer*.

4. At the authentication prompt, enter your *Apple ID* and *Apple ID password,* then tap *Deauthorize*.

5. Quit *Music*.

Sign out of iCloud

6. Select *Apple* menu > *System Settings* > *iCloud*.

7. Tap the *Sign Out* button.

8. At the prompt *Do you want to keep a copy of your iCloud data on this Mac before signing out?* This doesn't really matter, as you are erasing the drive. Tap *Keep a Copy.* Your iCloud data remains on iCloud, and on any other devices that are using the same Apple ID.

9. Quit *System Settings.*

Sign out of Messages

10. Open the *Messages* app, *Messages* menu > *Preferences.*

11. Select the *Messages* tab, then tap *Sign Out.*

12. Quit *Messages* app.

[Intel Mac] Reset NVRAM

Reset NVRAM erases some minor user settings from memory and restores a few security features to factory default.

13. Shut down your Mac.

14. Power on your Mac, then immediately press and hold *opt + cmd + P + R* key for 20 seconds.

[Optional] Unpair Bluetooth devices that you are keeping

15. While booted from your Mac, go to *Apple* menu > *System Settings* > *Bluetooth,* then tap the *remove (x)* button next to the device name.

Erase All Content and Settings

New with macOS 12 is the *Erase All Content and Settings* command.

• Prerequisite: An Intel Mac with a T2 security chip[2], or an Apple silicon Mac.

16. Log into your Mac with an administrator account.

17. Open *System Settings.*

18. Select *System Settings* menu > *Erase All Content and Settings* menu.

19. Enter your administrator password, then tap *OK.*

20. An alert will appear warning to back up to Time Machine.

------

[2] *https://support.apple.com/en-us/HT208862*

- If you do not have a current backup, tap *Open Time Machine* to start a new backup.

- If you do have a current backup, tap *Continue.*

21. The *Erase All Content & Settings* window appears. Tap the *Continue* button.

22. The *Sign Out of Your Apple ID* window appears. Enter your Apple ID password, then tap *Continue* button.

23. The 256-bit AES encryption key is deleted from the boot storage device and the Mac is rebooted. This effectively renders the computer securely erased, leaving only the existing macOS and standard applications installed, all user accounts, user files, and user-installed applications erased.

20.1.2 [Optional] [Intel Mac] Assignment: Remove Firmware Password

If you have created a firmware password, you must remove it prior to someone else using your computer.

In this assignment, you remove the Firmware Password. If you wish to leave it enabled, skip this assignment.

1. Shut down your Mac.

2. Boot into Recovery HD mode with these steps:

   a. Power on your Mac.

   b. Immediately after the startup tone, hold down the cmd + R keys.

   c. At the *Firmware Password* prompt, enter that password.

3. Select the *Utilities* menu > *Startup Security Utility.*

4. Select the *Remove Firmware Password* button.

5. Enter your *firmware password.*

6. Select the *Remove* button.

7. Select the *Apple* menu > *Restart* to restart the Mac.

20.2   Secure Erase Storage Devices

Before selling, giving away, or trashing your Mac, all data on the internal and external storage device(s) must be made inaccessible. There is only one safe option:

- Physically destroy the drive.

Most older desktops and laptops are equipped with the traditional hard disk drives (HDD). These drives have spinning magnetic platters that store data in a way that both the hard drive's controller and operating system can understand. When it comes to securely erasing a standard magnetic drive, all the individual magnetic regions are overwritten with random sequences of ones and zeros, effectively removing any "memory" they may have had prior to the procedure.

Solid State Drives (SSDs) operate a bit differently. Like their magnetic counterparts, SSDs store information as ones and zeros. But unlike magnetic disks, only the SSD's internal controller knows where in the memory cells the data is being stored. Because the SSD's controller uses wear leveling to maximize the life of the drive, the operating system has no idea where the controller ends up putting the ones and zeros.

A Hybrid or Fusion drive is not much more than a small SSD attached to an HDD.

When it comes time to securely erase your SSD, a wiping utility traditionally used for HDD may end up not evenly erasing a solid-state drive, leaving some data accessible. Because of this, it is important that the utilities you choose for wiping are specifically compatible with SSDs. These utilities use special SSD controller functions such as TRIM and Secure-Erase to sanitize your data.

I say we take off and nuke the entire site from orbit. It's the only way to be sure.

– Ellen Ripley[3]

---

[3] *https://en.wikipedia.org/wiki/Ellen_Ripley*

DoD, DoE, NSA, HIPAA, SEC Covered Organizations

If you must comply with DoD, DoE, NSA, HIPAA, SEC, or other top security regulations, you may have to physically destroy the drive of a computer that is being sold, given away, or disposed. This is because it is impossible to be 100% certain of 100% erasure of 100% of data.

If the computer is to be reused within the organization, a non-destructive alternative is to use *Parted Magic*[4]. This is Linux software that creates a bootable USB flash drive, which can then perform a secure erase on any storage device. The bootable USB flash drive works well with all PCs and Intel Macs. It is not certified yet for use on Apple silicon Macs.

Parted Magic cost is $11-$39 per year. You may purchase a license and download the software from *https://partedmagic.com*. Full step-by-step instructions may be found on the site.

Everyone Else

For the rest of us, if the computer is to be sold, given away, or disposed, I recommend following the same mandate for high security organizations–destroy the storage device. This is recommended because there is always the possibility your data may be recovered even if securely erased.

If the computer is to be reused within the business or household, using *Parted Magic* may be an acceptable alternative to storage device destruction.

MacOS has a built-in application to securely erase a storage device. *Disk Utility* provides the tools to erase HDD devices to Department of Defense standards, and hybrid/fusion and SSD devices adequate for the rest of us.

- HDD. If the storage device is not encrypted, the device may be securely erased by either of the following steps:

- Use *Disk Utility* to encrypt the device with a password, and then erase the device without encryption.

- Use *Disk Utility* to perform a multiple pass erase.

- HDD, fusion, and SSD. If the storage device has been encrypted using FileVault 2, use *Disk Utility* to perform an erase without encryption.

---

[4] *https://partedmagic.com/*

20.2.1 [Optional] Assignment: Secure Erase Boot Drive

If you are using a version of macOS prior to 12, there is no option to *Erase All Content and Settings.* In this assignment, you manually erase the computer boot drive.

- **Warning**: This assignment permanently erases all your data.
- Prerequisite: A full clone backup of the storage device to be erased.

1. If you will be securely erasing a boot drive, restart your computer into *Recovery HD mode* by holding down the *cmd+R* keys immediately after restart (for Intel Mac), or by pressing and holding the *Power* button (for Apple silicon Mac). Tap the *Options* icon, tap *Continue,* then select *Disk Utility.*

2. From the *Toolbar,* select *View* icon > *Show all devices.*

3. If you see any grayed volume names, select the volume name from the Toolbar select *Mount* icon, and if prompted, enter an administrator name and password.

4. From the sidebar, select the device to be erased. This is the left-most outdented name.

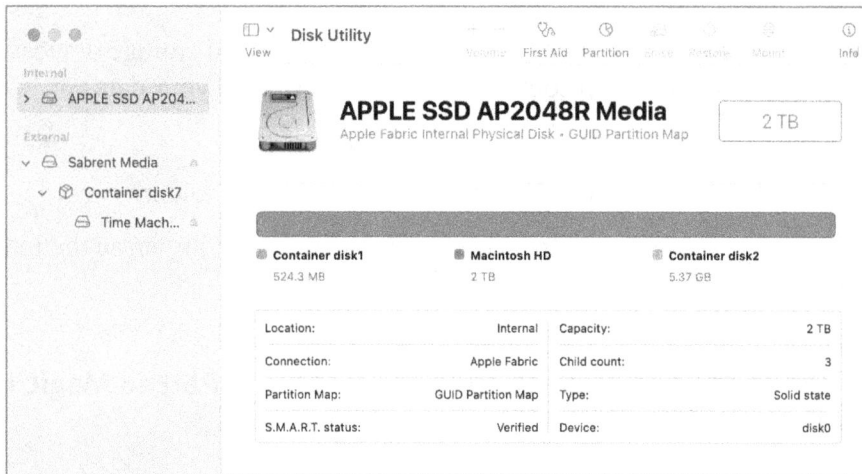

5. From the toolbar, select the *Erase* button.

6. From the *Erase* window:

a.   Enter a name for the volume. The default for a macOS boot drive is *Macintosh HD*.

b.   From the *Format:* pop-up menu, select *APFS* as the desired volume format.

c.   From the *Scheme:* pop-up menu, select *GUID Partition Map*.

7.   If the *Security Options...* button is available, tap it.

8.   Select the erase option best for your situation:

a.   The *Most Secure* (right most) option is the best, but it could take a day or more to complete.

b.   If time is of the essence, select the tick mark to the left of *Most Secure*.

c.   Then tap the *OK* button.

9.   Select the *Erase* button to start the erase process.

Done! The data on the drive is no longer readable.

20.2.2 [Optional] Assignment: Secure Erase a Non-Boot Drive

If you will be selling, giving away, or trashing your external storage devices, they also should be securely erased so nobody else has access to your data.

1.   Attach the external storage device to the computer.

2.   Open *Disk Utility,* located in the */Applications/Utilities* folder.

3.   Follow the same steps (as in previous assignment 20.2.1) to erase the internal storage device.

20.2.3 [Windows] Assignment: Create a Bootable Parted Magic USB Drive

20.2.4 [Windows] Assignment: Secure Erase An Encrypted SSD With Recovery Drive

20.2.5 [Windows] Assignment: Secure Erase an Unencrypted SSD With Parted Magic

20.2.6 [ChromeOS] Assignment: Secure Erase External Devices

20.3 Install macOS

It is not a matter of *if,* only a matter of *how often* you will need to reinstall macOS. The typical times this needs to be done are:

- You are selling or giving the computer to someone.

- The operating system is suspected of being damaged or corrupt.

- I recommend to all my clients to perform a *clean install* (full backup, reformat the boot drive, reinstall of macOS, reinstall applications, copy data from backup to boot drive) yearly. You will be amazed at how many little glitches are resolved and how much better your computer performs by doing this.

20.3.1 [Optional] Assignment: Install macOS 13

In this assignment, you install macOS 13.

If you are selling or giving this computer to someone, you can install a fresh version of macOS for them to easily get started on the new computer. If you are discarding the computer, you may skip this assignment.

- **Warning**: Performing this assignment will permanently delete everything on your boot drive, then reinstall macOS 13. Perform this assignment only if you have a current backup and want to wipe your drive.

1. Boot into Internet Recovery HD mode. The steps are:

   o Intel Mac: Power on, then immediately hold down the *opt + cmd +R* keys and continue holding down until the Apple logo appears.

   o Apple silicon Mac: Press and hold the *Power* button until *the Start* Manager screen appears (you will see the boot drive and *Options* icons).

415

2. Tap on the icon for your user account, then tap *Continue*.

3. Enter your login password, then tap *Continue*.

4. At the *macOS Utilities* window, tap *Reinstall macOS*, then tap *Continue*.

5. Follow the on-screen instructions to complete installation.

6. When installation completes, the Mac restarts.

7. At the *Setup Assistant* window:

    a. If you are keeping the Mac, continue following the on-screen instructions.

    b. If you are selling or giving away the Mac, press *Command-Q* to quit out of the assistant, then tap *Shut Down.* When the new owner powers on the Mac, it boots to the *Setup Assistant,* allowing them to customize the setup.

20.3.2 [iOS] Assignment: Prepare a Device for Sale or Disposal

20.4   When it is Time to Say Goodbye Lessons Learned

☐ Before selling, giving away, or even trashing a device, it must be properly processed.

☐ The process to say goodbye to a Macintosh is to: create a backup, sign the device out of Music, sign the device out of iCloud, sign the device out of Messages, reset NVRAM (Intel Mac), unpair Bluetooth devices, and erase the storage device.

☐ New with macOS 12 is the *Erase All Content and Settings* command, available within System Settings. This effectively securely erases the boot drive by erasing the AES 256-bit encryption key. This works on Apple silicon Macs and Macs with a T2 security chip.

21 Miscellaneous

The nice thing about standards is that you have so many to choose from.

–Andrew S. Tanenbaum[1]

What You Will Learn in This Chapter

- Configure date and time settings
- Comply with National Institute of Standards and Technology (NIST)
- Comply with United States Computer Emergency Readiness Team (US-CERT)
- Comply with International Organization for Standardization (ISO)

What You Will Need in This Chapter

- No additional resources required.

21.1   Date and Time Settings

There are several reasons it is critical to keep your computer date and time accurate:

- If you are on a network with other computers, or use a server, if your clock is off by more than a few minutes, you may be blocked as the other systems see this as a potential *man in the middle* attack.

---

[1] *https://en.wikipedia.org/wiki/Andrew_S._Tanenbaum*

- Should your device become compromised by malware or a criminal, it is important to know the exact moment the penetration occurred.

- You do not want to be late to your Aunt Rose's dinner party. Noodle Koogle is best right out of the oven.

Fortunately, most devices with internet connectivity include the ability to automatically synchronize with atomic clocks around the world using the Network Time Protocol[2] (NTP).

21.1.1 Assignment: Configure Date & Time

In this assignment, you configure your device to use the Apple NTP server.

1. Open *Apple* menu > *System Settings* > *General* > *Date & Time*, then configure to your taste. My settings are below:

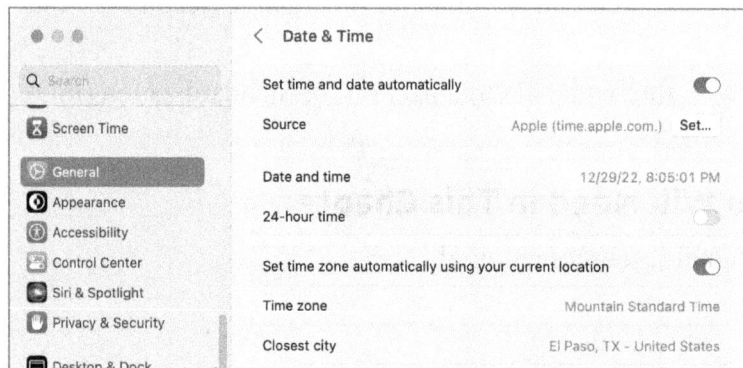

a. Enable the checkbox *Set date and time automatically*.

b. In the *Source* field, almost everyone can leave set to the default of *Apple (time.apple.com)*. If you need to set to a different NTP server (such as in an office), tap the *Set* button, then enter the FQDN or IP address of the target server.

c. Disable *Set time zone automatically using your current location*.

d. In the Closest *City* field, tap the *Set* button, enter the name of the closest city, and then tap *Done*.

---

[2] *https://en.wikipedia.org/wiki/Network_Time_Protocol*

2. Close System Settings

Your computer now automatically synchronizes its time with the configured NTP server.

21.2 National Institute of Standards and Technology (NIST)

NIST, part of the U.S. Department of Commerce, was established by Congress in 1901 to improve our measurement systems because substandard measurement systems were impeding U.S. economic growth. This NIST mission continues into the 21st-century by developing standards of measurement at the nanoscale through galactic scale.

NIST also is involved in developing standards for computer and IT systems. Some organizations–particularly healthcare, financial, and legal–base much of their cybersecurity measures on the *NIST SP 800-171 Protecting Controlled Unclassified Information in Nonfederal Systems and Organizations*[3].

As of the time of this writing, NIST had just released *Draft NIST Special Publication 800-179 Guide to Securing Apple macOS 10.12 Systems for IT Professionals: A NIST Security Configuration Checklist*[4]. This publication details how NIST recommends configuring the security of macOS 10.12.

NIST also publishes the *NIST SP 800-53 Security and Privacy Controls for Information Systems and Organizations*[5]. The current release is revision 5, released September 2020. This document was originally intended for standards and guidance for federal organizations. However, with the most current release, the name "Federal" was removed from the title in recognition that the guidelines are appropriate for all organizations.

---

[3] *https://nvlpubs.nist.gov/nistpubs/SpecialPublications/NIST.SP.800-171r1.pdf*
[4] *https://csrc.nist.gov/CSRC/media/Publications/sp/800-179/rev-1/draft/documents/sp800-179r1-draft.pdf*
[5] *https://csrc.nist.gov/publications/detail/sp/800-53/rev-5/final*

The *macOS Security Compliance Project*[6] has put the 800-53 guidelines into a somewhat more human-readable format for macOS users, to be found here[7]

Following the steps outlined in *Practical Paranoia*, you can likely pass a NIST-based security audit. That said, if your organization is held to NIST protocols, you are held accountable for knowing them from the NIST perspective, and reading their guide is mandatory.

21.2.1 [Optional] Assignment: NIST Department of Defense Configuration

NIST maintains Department of Defense IT checklists[8] for operating systems, servers, and applications. If your organization must comply with DoD protocols, this is a perfect resource.

21.2.2 [Android and iOS] Assignment: NIST-Specific Electronic Health Records on Mobile Devices

21.3 United States Computer Emergency Readiness Team (US-CERT)

US-CERT[9], part of the Department of Homeland Security, began in early 2000 as a response to Federal Government networks experiencing an alarming number of cyber breaches. Congress created the Federal Computer Incident Response Center (FedCIRC) at the General Services Administration as a centralized hub of coordination and information sharing between federal organizations. With the creation of the Department of Homeland Security in 2002, Congress transferred US-CERT responsibilities to FedCIRC. In 2003, FedCIRC was renamed US-

---

[6] *https://support.apple.com/guide/sccc/macos-security-compliance-project-sccc22685bb2/1/web/1.0*

[7] *https://github.com/usnistgov/macos_security/releases/tag/big_sur_rev2*

[8] *https://nvd.nist.gov/ncp/repository?sortBy=modifiedDate%7Cdesc*

[9] *https://www.us-cert.gov/*

CERT, and its mission was expanded to include providing boundary protection for the federal civilian executive domain and cybersecurity leadership.

This shared responsibility has evolved over time to make US-CERT a trusted partner and authoritative source in cyberspace for the Federal Government, State, Local, Tribal and Territorial governments, private industry, and international organizations.

21.4   International Organization for Standardization (ISO)

ISO[10] is an independent, international, standards organization. Many industries–in particular those dealing with banking, SEC-compliance, and financial advisors–rely on one of the many ISO standards. The ISO 27001[11] is the primary standard covering IT security.

Although many books are available to assist with meeting ISO 27001 compliance, it is best to first purchase the ISO 27001 book[12], which details the compliance matrix. This is only available from the ISO.

21.5   Vehicle Infotainment Systems

In August 2016, the FTC[13] and US-CERT[14] released an alert regarding securing personal information when using rental vehicles.

When using a rental vehicle with smart features, it is likely these tools are storing data about your travel, including full GPS mapping of your path, your speed at each point along the path, where you stopped, and for how long. Unfortunately, it

---

[10] *https://www.iso.org*

[11] *https://www.iso.org/isoiec-27001-information-security.html*

[12] *https://www.iso.org/standard/54534.html*

[13] *https://www.consumer.ftc.gov/blog/what-your-phone-telling-your-rental-car*

[14] *https://www.us-cert.gov/ncas/current-activity/2016/08/30/FTC-Releases-Alert-Securing-Personal-Information-When-Using-Rental*

is unlikely you can do much about this as most of the systems do not allow erasing this data.

When connecting a digital device such as a smartphone or tablet to a vehicle, it is possible the vehicle infotainment system automatically imports personal data such as contacts, your phone number, text messages and emails.

This information is automatically saved in the infotainment memory, and may be viewed by the rental agency personnel, or the next renter. Some of this information may be used for identity theft, or other criminal activity.

To secure your personal data in a rental vehicle:

- **Do not connect your digital device to the infotainment system.** Instead of connecting to the infotainment USB port, connect through a cigarette lighter USB adapter for charging. Some USB connections automatically transfer data upon connection.

- **Configure permissions.** If you do want or need to connect your mobile device to the infotainment system, the system may present a permissions screen to specify or limit what data is transferred. If so, transfer only your music list, not contacts, messages, emails, etc.

- **Delete all data upon returning the rental vehicle.** Prior to returning the rental vehicle, check the infotainment system settings for the list of devices it is paired with. Locate your device and delete it from the list.

21.6 [Android and iOS] Emergency Location Service

21.7 [iOS] Strong Hacking Protection

21.8 Miscellaneous Lessons Learned

☐ Maintaining accurate time on your device is critical for security.

☐ macOS devices have the built-in ability to synchronize with the atomic clocks participating in the Network Time Protocol (NTP), as well as local network servers.

☐ National Institute of Standards and Technology (NIST), part of the US Department of Commerce, is a major researcher and developer of cybersecurity and privacy best practices.

☐ United States Computer Emergency Readiness Team (US-CERT), part of the Department of Homeland Security, is tasked with developing standards of cybersecurity for federal government agencies, as well as non-governmental organizations that work with federal agencies.

☐ International Organization for Standardization (ISO), an independent international organization, creates standards for many industries including financial.

☐ When using a rented vehicle, do not connect your digital devices to the infotainment system. Instead, connect to a cigarette lighter USB adapter for charging.

22 The Final Word

If you followed each of the steps outlined in this book, your computer is secured to a level the NSA requires for most of its own staff. Although this will not prevent all the bad guys from stealing your computer, it prevents them from accessing your data. And since you have at least one current backup at the home or office, and one on the Internet, you still are in possession of the items with *real* value: your data and peace of mind.

22.1   Additional Reading

"Security Recommendations to Prevent Cyber Intruders." US-CERT Alert (TA11-200A). July 2011. US-CERT. <https://www.us-cert.gov/ncas/alerts/TA11-200A>

23 [ChromeOS] Linux

macOS 13 Security Checklist

Included below is the checklist used by Mintz InfoTech, Inc. consultants when performing security checks for our clients' systems. You should use this same checklist to ensure your own system is fully hardened.

OS Version

- ☐ Verify OS is up to date (1.14.1)

Data Loss

- ☐ Time Machine is active and encrypted, with password securely recorded (2.1.2, 2.1.3)

- ☐ Internet-based backup daily backup, password securely recorded (2.1.10, 2.1.15, 2.1.16)

- ☐ Integrity test Time Machine backup monthly (2.1.3)

- ☐ Integrity test Internet-based backup monthly (2.1.14)

Passwords

- ☐ All passwords are strong and securely recorded (3.2.1, 3.2.2)

- ☐ All challenge questions and answers securely recorded (3.4.1)

- ☐ 2-Factor Authentication in use for all available sites and services, using either a 2FA app or 2FA USB key (3.8-3.8.2)

- ☐ Synchronize passwords across devices and browsers with Bitwarden or iCloud Keychain (Note: Not for use in secure facilities) (3.5-3.6.2)

Optional

- ☐ Harden the password manager with a timed lock (3.3.4)

- ☐ Assign Password Policies (3.7-3.7.2)

System and Application Updates

- ☐ Verify OS is up to date and automatically updates (4-4.1.1)

☐ Verify applications are up to date and automatically updated (4.1.1, 4.1.2, 4.1.3)

☐ Configure System and application privacy (4.2-4.2.2)

☐ Verify Gatekeeper configuration (4.2.2)

Optional

☐ Install MacUpdater to automate application updates (4.1.3)

User Accounts

☐ All users log in with non-administrative accounts (5.2-5.2.6)

☐ Root user not enabled (5.2.4)

☐ Enable a Policy Banner (5.4, 5.4.1)

Optional

☐ Screen Time enabled (5.3-5.3.2)

☐ Screen Time Whitelisting enabled (5.3-5.3.2)

Device Hardware

☐ Enable FileVault 2 Full Disk Encryption for all storage devices, from boot drive to thumb drives (6.2-6.2.8)

☐ Enable Firmware password (Intel Macs) (6.5-6.5.2)

☐ Configure Startup Security (Apple Silicon Macs) (6.5.4)

☐ User is aware of microphone and camera recording indicator (6.8-6.8.2)

Sleep and Screen Saver

☐ Sleep or screen saver set to start in 5-15 minutes of inactivity (7.1.1)

☐ Screen notifications restricted on screen lock (7.2.1)

Optional

☐ Do Not Disturb configured (7.3.1)

Malware

☐ Quality antimalware installed and up to date (8-8.1.2)

☐ After antimalware installation, perform a full scan of all drives (8.1.1, 8.1.2)

☐ Verify antimalware updates daily (8.1.1, 8.1.2)

Optional

☐ Enable Lockdown Mode (8.6)

Firewall

☐ Firewall is enabled (9.1.1)

☐ All unnecessary ports closed (9.1.2)

Lost or Stolen Device

☐ Verify Find My Mac active (10.1.1)

☐ Enable Energy Saver Wake for Network Access (Note: Not for use in secure facilities) (10.1.1)

☐ Enable Security & Privacy Location Services (10.1.1)

Local Network and Bluetooth

☐ No Ethernet hubs in use, only Ethernet switches (11.4)

☐ WPA3 with AES active for all Wi-Fi networks (11.4.1, 11.4.2)

☐ Strong password in use for Wi-Fi (11.4.2)

☐ WPA3 in use with router (11.4.2)

☐ Network passwords securely recorded

☐ Power-cycle modems and routers (11.6.1)

☐ Update firmware for modems and routers (11.6.1)

☐ Verify DNS settings on modems and routers (11.6.1)

☐ Verify Port Forwarding settings on modems and routers (11.6.1)

☐ Verify DMZ settings on modems and routers (11.6.1)

☐ Pair Bluetooth devices in a secure area, away from possible eavesdroppers (11.7.1, 11.7.2)

☐ Do not authorize Bluetooth connections unless you are trying to pair it for the first time (11.7.2)

☐ Remove lost Bluetooth devices from the paired device list (11.7.3)

Optional

☐ Enable MAC Address filtering to Limit Wi-Fi Access (11.5.1)

☐ Install RADIUS authentication for Wi-Fi and Ethernet access (11.1)

Web Browsing

☐ Never enter sensitive information in an HTTP page, only on an HTTPS page (12.1)

☐ High privacy browser in use (12.2)

☐ DuckDuckGo is the default search engine (12.4.1)

☐ Browser Do Not Track enabled (12.8)

☐ Browser Block 3rd-party Cookies enabled (12.2.1, 12.2.2)

☐ Browser Fraudulent Website Warning enabled (12.2.1, 12.2.2)

☐ Check for account breaches at haveibeenpwned.com (12.12.1)

☐ Update browser extensions and remove unneeded or suspicious extensions (12.6.1, 12.6.2)

☐ Apple Personalized Ad Tracking disabled (12.13.1)

☐ Review Browser security settings 12.2.1, 12.2.2)

☐ Block Fingerprinting enabled (12.8)

Optional

☐ Install and configure Brave browser (12.2.1)

☐ Enable Private Browsing (12.3)

☐ Install and configure Tor (12.10.1-12.10.3)

Email

☐ User educated on how to recognize phishing attacks (13.2)

☐ TLS or HTTPS used for all email (13.4-13.5.1)

- ☐ Mail Privacy Protection enabled for @mac.com and @icloud.com accounts (13.6-13.6.1)
- ☐ Create SPF, DKIM, and DMARC records (13.11-13.11.4)
- ☐ Enable 2-Factor Authentication (13.12)

Optional

- ☐ Email aliases used for subscriptions (13.7-13.7.3)
- ☐ Outlook.com used for encryption and prevent forwarding (13.9-13.9.2)
- ☐ ProtonMail account used (13.8-13.8.3)
- ☐ Paubox account used (13.4.5)

Documents

- ☐ Know how to encrypt MS Office documents (14.2-14.2.1)
- ☐ Know how to encrypt a PDF document (14.3-14.3.1)
- ☐ Know how to encrypt a folder for macOS use (14.4-14.4.1)
- ☐ Know how to encrypt a folder for cross platform use with Zip (14.5-14.5.3)
- ☐ Know how to remove sensitive Exif data from images and audio files (14.6-14.6.2)
- ☐ Know how to remove sensitive metadata from files (14.6.3-14.6.8)

Optional

- ☐ Install Keka (14.5-14.5.3)

Voice, Video, And Instant Message Communications

- ☐ Signal installed and used for secure voice, video, and text communications (15.1-15.3.4)

Internet Activity

- ☐ A quality VPN is always in use (16-16.5)
- ☐ Secure DNS is always in use (16.6-16.6.2)
- ☐ Perform DNS leak tests monthly (16.6.1)

☐ Private Relay is in use (16.7.1)

Social Media, Apple ID and iCloud

☐ Strong, unique passwords used for each social media site (17.3.1)

☐ All social media passwords securely recorded

☐ Enable 2-Factor Authentication for all available sites (17.3.1)

☐ Review all site privacy and ad settings, and apps

☐ Review profile as a friend, connection, etc. to verify no problem content

☐ Verify all devices that can receive 2FA codes

☐ Review Off-Facebook data (17.3.3)

☐ Review what LinkedIn knows about you (17.4.2)

☐ Download Google Takeout data (17.5.2)

☐ Enable Microsoft 2-Factor Authentication (17.6.1)

☐ Remove any device from your Microsoft account that is not currently used by you (17.6.2)

☐ Enable Apple 2-Factor Authentication (17.7.1)

☐ Remove any device from your Apple account that is not currently used by you (17.7.2)

☐ Verify Apple Account Recovery configuration (17.8.1)

☐ Verify Apple Legacy Account configuration (17.8.1)

Apple Pay and Credit Cards

☐ Verify device Apple Pay configuration (18.1.1)

☐ Know how to use Apple Pay with online stores (18.1.2)

When it is Time to Say Goodbye

☐ Create a local full backup (20.1.1)

☐ Sign out of Music/iTunes (20.1.1)

☐ Sign out of iCloud (20.1.1)

☐ Sign out of Messages (20.1.1)

☐ Reset NVRAM (Intel Mac) (20.1.1)

☐ Unpair Bluetooth devices (20.1.1)

☐ Secure erase all contents and settings (20.1.1, 20.2.1)

☐ Secure erase storage device (20.2, 20.2.2, 20.2.3)

☐ Remove Firmware password (Intel Mac) (20.1.2)

Optional

☐ Fresh install of macOS (20.3.1)

Miscellaneous

☐ Configure Date & Time to synchronize with an NTP server (21.1-21.1.1)

☐ Rigorously read and follow the NIST 800-171 protocols if working in Federal Contractor, legal, healthcare, or financial environment (21.2-21.2.1)

Revision Log

6 20230114

- Chapter and topic alignment with Practical Paranoia ChromeOS.
- Minor edits.

6 20221224

- Fixed a production error that substituted macOS 12 sections for macOS 13.

6 20221219

- Expanded topic alignment with Practical Paranoia Android 11 and higher.

6 20221214

- Added 17.9 Advanced Data Protection.

6 20221022

- Initial release, fully updated and rewritten for macOS 13.
- Chapter and topic alignment with Practical Paranoia Android 12, ChromeBook, iOS 16, and Windows 11.

Acronyms

2FA–*Two-Factor Authentication*. The use of a secondary source (SMS message, email, authenticator app, authenticator key) for authentication in addition of username and password.

ACSP–*Apple Certified Support Professional*. An Apple certification required as part of acceptance into the Apple Consultants Network.

ACTC–*Apple Certified Technical Coordinator*. Currently, the highest-level Apple certification.

AES–*Advanced Encryption Standard*. An algorithm for the encryption of electronic data.

BAA–*Business Associate Agreement*. A US legal document required of some persons and organizations who perform services for HIPAA-covered entities (health organizations).

CEO–*Chief Executive Officer*. The person who thinks they are the highest-ranking in an organization.

CIO–*Chief Information Officer*. The person who actually controls an organization, even if they typically report to the CEO.

CISPA–*Cyber Intelligence Sharing and Protection Act*. Allows sharing of internet traffic between the US Government and technology and manufacturing organizations.

DKIM–*DomainKeys Identified Mail*. An email authentication method designed to detect forged sender addresses.

DMARC–*Domain-based Message Authentication, Reporting, and Conformance*. An email authentication protocol, designed to give domain owners the ability to protect their domain from unauthorized use.

DMZ–*Demilitarized Zone*. A physical or logical subnetwork that contains all the external-facing services to an untrusted network, typically the internet.

DNS–*Domain Name Server* or *Domain Name System*. A hierarchical and decentralized naming system for computers, services, and other resources connected to a local network or internet.

DSL–*Digital Subscriber Line*. A technology for high-speed digital data over standard phone lines.

Exif–*Exchangeable image file format*. A standard for storing metadata for image and sound files.

FTC–*Federal Trade Commission* (US). A US antitrust and consumer protection agency.

GB–*Gigabyte*. A unit of information equal to 2^{30} bytes or approximately one billion bytes.

GPG–*GNU Privacy Guard*. Free cryptographic software.

HDD– *Hard Disk Drive*. A type of storage device containing rigid rotating platters.

HIPAA–*Health Insurance Portability and Accountability Act*. A US law that restricts access to individuals' private medical information.

HTTP–*Hypertext Transport (or Transfer) Protocol*. The data transfer protocol used on the World Wide Web.

HTTPS–*Hypertext Transfer Protocol Secure*. The encrypted data transfer protocol used on the World Wide Web.

iOS–The name of the operating system used by Apple iPhones.

IoT–*Internet of Things*. The network of objects connected to the internet, not normally viewed as computers or telecommunications. Examples are nanny cams, smart doorbells, etc.

IP–*Internet Protocol*. A standardized protocol for the transmission of digital data over both local networks and the internet.

iPadOS–The name of the operating system used by Apple iPads.

ISP–*Internet Service Provider*. A company that provides subscribers with access to the internet.

ISO–*International Organization for Standardization*. An international standard-setting body composed of representatives from various national standards organizations.

IT–*Information Technology*. The study or use of computers or telecommunications for storing, retrieving, and sending information.

LED–*Light Emitting Diode*. A semiconductor which glows when voltage is applied.

MAC–*Media Access Control address*. A unique identifier assigned to a network interface controller, typically used with computers, printers, networking devices, mobile phones, and any device that connects to a local network or internet.

macOS–The name of the operating system used by Apple Macintosh computers.

MB–*Megabyte*. A unit of information equal to 2^{20} bytes or approximately one million bytes.

MBA-IT–*Master of Business Administration, Information Technology specialization*. A graduate-level university program and degree designed for individuals seeking technology leadership positions.

MP–*Military Police*.

MX–*Mail Exchanger record*. Specifies the mail server responsible for accepting email messages on behalf of a domain name.

NIST–*National Institute of Standards and Technology* (USA-based). A physical sciences laboratory and non-regulatory agency of the United States Department of Commerce. Mission is to promote innovation and industrial competitiveness.

OMG–*Oh My God!*

OS–*Operating System*. The code running at the heart of computers.

PDF–*Portable Document Format*. A file format designed to include text, images, sound, video, and other media, in a manner independent of application software, hardware, and operating system.

PGP–*Pretty Good Privacy*. Cryptographic software.

PIN–*Personal Identification Number*. Typically used to validate electronic transactions.

PRISM–Code name for a program under which the United States National Security Agency collects internet communications from US internet companies.

QR Code–*Quick Response Code*. A machine-readable code consisting of an array of black and white squares. Typically used to store URLs and other information for reading by the camera on a smartphone.

RADIUS–*Remote Authentication Dial-In User Service*. A networking protocol providing centralized authentication, authorization, and accounting management for users connecting to a network service.

RAM–*Random Access Memory*. This is the area holding data as it is needed by the computer.

SIM–*Subscriber Identification Module*. A smart card inside a mobile phone, carrying an identification number unique to the owner, storing personal data, and preventing operation if removed.

S/MIME–*Secure/Multipurpose Internet Mail Extensions*. A standard for public key encryption and signing of MIME data.

SPF–*Sender Policy Framework*. An email validation system.

SOS–An international code signal of extreme distress.

SSD–*Solid-State Drive*. A storage device that uses integrated circuit assemblies to store data persistently.

SSL–*Secure Sockets Layer*. A digital encryption protocol.

TB–*Terabyte*. A unit of information equal to 2^{40} bytes or approximately one million million bytes.

TCP–*Transmission Control Protocol*. A set of rules for the delivery of data over a local network or internet.

TKIP–*Temporal Key Integrity Protocol*. An algorithm used to secure wireless networks.

TLS–*Transport Layer Security*. A digital encryption protocol.

Tor–*The Onion Router*. A highly secure web browser which uses both data encryption and random bouncing between multiple routers to help ensure security and privacy.

TXT–*Text record.* A type of resource record in the DNS used to provide the ability to associate arbitrary text with a host or other name, such as human readable information about a server, network, or data center.

US-CERT–*United States Computer Emergency Readiness Team.* Organized within the Department of Homeland Security's Cyber Security and Infrastructure Security Agency.

VPN–*Virtual Private Network.* A protocol/software that allows for end-to-end encrypted communication or data transmission over a local network or internet.

WEP–*Wired Equivalency Protocol.* An early encrypted wireless networking protocol.

WPA–*Wi-Fi Protected Access.* As of version 3, the current encrypted wireless networking protocol.

Index

www.ingramcontent.com/pod-product-compliance
Lightning Source LLC
Chambersburg PA
CBHW080133220326
41598CB00032B/5054